QUANTUM THEORY, BLACK HOLES AND INFLATION

QUANTUM THEORY, BLACK HOLES AND INFLATION

Ian G. Moss
University of Newcastle upon Tyne

JOHN WILEY & SONS
Chichester · New York · Brisbane · Toronto · Singapore

Other Wiley Editorial Offices

John Wiley and Sons, Inc., 605 Third Avenue,
New York, NY 10158-0012, USA

Jacaranda Wiley Ltd, 33 Park Road, Milton,
Queensland 4064, Australia

John Wiley & Sons (Canada) Ltd, 22 Worcester Road,
Rexdale, Ontario M9W 1L1, Canada

John Wiley & Sons (SEA) Pte Ltd, 37 Jalan Pemimpin #05-04,
Block B, Union Industrial Building, Singapore 2057

Library of Congress Cataloging-in-Publication Data:
Moss, Ian G.
    p.   cm.
    Quantum theory, black holes, and inflation / Ian G Moss.
    Includes bibliographical references and index.
    ISBN 0–471–95736–4 (alk. paper)
    1. Cosmology. 2. Black holes (Astronomy) 3. Quantum theory.
    4. Inflationary universe. I. Title.
QB981. M68   1996
523.1—dc20   95–25011
                                                  CIP

British Library Cataloguing in Publication Data:
A catalogue record for this book is available from the British Library

ISBN 0 471 95736 4

Produced from camera-ready copy supplied by the author
Printed and bound in Great Britain by Biddles Ltd, Guildford, Surrey
This book is printed on acid-free paper responsibly manufactured from sustainable forestation, for
which at least two trees are planted for each one used for paper production

В ОЛЬГА

Contents

Notation and conventions

In this book physical quantities are expressed in Planck units, constructed from Newton's constant G, reduced Planck constant $\hbar$ and the speed of light c:

length unit L_p	$(\hbar G/c^3)^{1/2}$	$1.616 \times 10^{-35}\,\mathrm{m}$
mass unit M_p	$(\hbar c/G)^{1/2}$	$2.177 \times 10^{-8}\,\mathrm{kg}$
time unit T_p	$(\hbar G/c^5)^{1/2}$	$5.391 \times 10^{-44}\,\mathrm{s}$
energy unit E_p	$(\hbar c^3/G)^{1/2}$	$1.221 \times 10^{19}\,\mathrm{GeV}$

The fundamental constants can always be restored to a formula by dividing the terms by an appropriate unit factor from the table. As an aid to checking formulae, Newton's constant and the reduced Planck's constant have been retained wherever practicable.

This book is aimed at readers who are already familiar with the basic framework of general relativity. Bold-face characters will generally be used in formulae to denote vectors or tensors and three-vectors are bold italic, e.g. $\boldsymbol{x}$. Coordinate axes are labelled by Greek indices which run from $0, \ldots, 3$.

Lorentz metrics have signature $(-+++)$ here and Riemannian metrics have signature $(++++)$. Lie derivatives $\mathcal{L}_{\mathbf{X}}$ and space-time covariant derivatives $\nabla_{\mathbf{X}}$ of a tensor field $\mathbf{T}$ along a vector field $\mathbf{X}$ adopt the following pattern:

$$(\mathcal{L}_{\mathbf{X}}\mathbf{T})^{\mu}{}_{\nu} = X^{\rho}\frac{\partial}{\partial x^{\rho}}T^{\mu}{}_{\nu} - \frac{\partial X^{\mu}}{\partial x^{\rho}}T^{\rho}{}_{\nu} + \frac{\partial X^{\rho}}{\partial x^{\nu}}T^{\mu}{}_{\rho}$$

$$(\nabla_{\mathbf{X}}\mathbf{T})^{\mu}{}_{\nu} = X^{\rho}\left(\frac{\partial}{\partial x^{\rho}}T^{\mu}{}_{\nu} + \Gamma^{\mu}{}_{\sigma\rho}T^{\sigma}{}_{\nu} - \Gamma^{\sigma}{}_{\mu\rho}T^{\mu}{}_{\sigma}\right).$$

Conventions for the connection $\Gamma^{a}{}_{bc}$ and the Riemann curvature tensor $R^{a}{}_{bcd}$ follow Misner et al. (1973).

Some of the quantities that appear in more than one chapter are listed below:

$\mathcal{A}$	Black hole surface area.
κ_h	Black hole surface gravity.
$\mathcal{S}$	Entropy.
Λ	Cosmological constant.
H	Hubble constant.
Ω	Density parameter.
h	Hubble parameter $H/(100\,\mathrm{km\,s^{-1}\,MPc^{-1}})$
∇	Spacetime covariant derivative.
$\mathbf{D}$	Gauge covariant derivative.
R	Ricci scalar.
$d\mu$	Volume measure on a manifold.
$d\mu[x]$	Measure on a space of functions $x(t)$.
$\mathcal{P}(\mathcal{C})$	Selection function for conditions $\mathcal{C}$.
$\mathcal{P}(x,t)$	$\mathcal{C} = $ 'passes through the point (x,t)'.
S	Lorentzian action.
I	Riemannian action.
I_J	Riemannian action with source J.
$\zeta(s)$	Generalised zeta-function.

1

Introduction

How did the universe begin? What happens at the centre of a black hole? These are issues balanced on the interface between quantum theory and gravitation. No-one has yet found convincing answers to these questions, but there are many results which present some interesting pictures of how these questions might begin to be answered.

This book focuses on results which retain most of the basic structure of general relativity, namely a manifold of spacetime points and a metric. Within this framework quantum theory introduces the possibility of quantum tunnelling and some of the most radical differences between the classical and quantum theory, including black hole evaporation.

1.1   Black hole evaporation

The prediction of black hole evaporation by Stephen Hawking in 1974 is a very important starting point for many of the ideas in this subject. The influence of this result can hardly be overstated, in particular the stimulus that this gave to the study of quantum fluctuations in various spacetimes. One consequence of this work is the exciting possibility that quantum fluctuations analogous to those responsible for black hole evaporation may be the origin of temperature fluctuations seen in the cosmic microwave background.

In order to set the scene, Bardeen, Carter and Hawking had, by the early 1970s, arrived at a number of important results about the nature and properties of black holes. These can be summarised in the four laws of black hole mechanics:

1. A quantity called the surface gravity κ_h is constant over the surface of the black hole.
2. The change in mass due to external actions is given by

$$dM = \frac{\kappa_h}{8\pi G}d\mathcal{A} + \text{work terms} \qquad (1.1)$$

   where $\mathcal{A}$ is the area.

3. The area of a black hole always increases

$$d\mathcal{A} \geq 0. \tag{1.2}$$

4. The surface gravity cannot be reduced to zero.

In order to qualify as a black hole the surface has to divide spacetime into two regions as it evolves in time. Light from the outer region can escape to infinity but the light from the inner region is trapped. The event horizon between the two regions is generated by a special set of null lines or light rays whose curvature determines the surface gravity κ_h. For a Schwarzschild black hole the surface gravity is $\kappa_h = 1/(4GM)$.

Quantities that appear in these laws are meant to be measured during periods when the spacetime is very nearly stationary. Fortunately, both perturbation theory and numerical calculations have shown that when black holes are left alone for a period of time they do generally settle down and approach a stationary state.

The laws of black hole mechanics certainly suggest that entropy can be attributed to black holes, with $\mathcal{S} \propto \mathcal{A}$, $T \propto \kappa$ and the energy $E = M$. This would also be consistent with the idea that information is lost during the formation of a black hole, as was first pointed out by Beckenstein (1974). However, the laws can only be interpreted in this way if a black hole exchanging energy with other systems can come to thermodynamic equilibrium. This generally fails for a classical black hole. With a bath of radiation, for example, the black hole can only absorb and can never emit radiation.

We now know that black holes do radiate particles due to quantum effects. In fact the temperature is given by

$$T = \frac{\hbar}{2\pi}\kappa_h. \tag{1.3}$$

From the first law we deduce that

$$\mathcal{S} = \frac{1}{4G\hbar}\mathcal{A}. \tag{1.4}$$

The detailed results are presented in chapter 7. For the present we can simply picture black hole evaporation as being due to quantum fluctuations of the vacuum near to the black hole horizon. In the regular quantum state, the flux of ingoing particle fluctuations does not balance the outward-going flux and there is a net thermal flow of particles from the hole. Overall energy conservation implies that the hole loses mass, and as it does so it becomes hotter.

The quantity of radiation from a black hole as massive as the Sun, for example, is quite negligible. The radiation is only likely to have astrophysical significance for small black holes, such as might be produced in the early stages of the universe. These primordial black holes would radiate into the X-ray background. This enables us to set stringent limits on their number, particularly for masses $M \leq 10^{15}$ g, whose lifetime is less than the age of the universe.

The analysis of these quantum fluctuations has led to the discovery of the importance of imaginary ($i = \sqrt{-1}$) time. Quantum statistical mechanics in flat space can also be formulated with imaginary time, but the black hole spacetime has very special topological features. Only when the imaginary time coordinate is identified with period $2\pi/\kappa_h$ does the metric have a regular extension to the horizon. It is this that fixes the temperature of the hole.

Perhaps one of the influential developments of black hole evaporation will be to advance our understanding of quantum gravity. At present, the treatment of black hole evaporation is incomplete because the approximations which are used break down when the temperature reaches one Planck unit, around 10^{19} GeV. Beyond this we can only guess. One possibility is that the radiation stops and a new particle is left behind. This superheavy particle would carry all of the quantum numbers of the original black hole and could be important in astrophysics and cosmology.

Another possibility is that the black hole evaporates away completely. This seemingly innocent proposal has some very interesting consequences. There would be a violation of some charge conservation laws. The number of baryons is an example, because a black hole formed from baryons alone will radiate equal amounts of baryons and antibaryons, leaving behind a net baryon number zero.

Another issue arises if the theory of quantum gravity somehow prevents singularities, raising the question of what happens to particles which propagate into the hole before it evaporates away? One possible reply is that a disconnected 'baby universe' forms to accommodate those particles which fell into the black hole (Hawking 1987). However, spacetime manifolds with Lorentzian metrics and a well defined ordering of events with respect to time cannot develop baby universes.

Ideas like these lead naturally to the Euclidean quantum gravity programme, where metrics with signature $(++++)$ play a predominant role. This is the signature of the analytic continuation of the black hole metric to imaginary time. It is also the most natural signature to use for describing the formation of baby universes.

The original name for this programme has the drawback that mathematicians usually reserve the adjective 'Euclidean' for spaces that are also flat. In order to avoid any possibility of confusion, in this book the adjective 'Riemannian' will be used for metric signature $(++++)$ and 'Lorenzian' for signature $(-+++)$.

1.2 Relativistic quantum theory

The theory of black hole evaporation introduces an important subject that will be developed in chapters 2 and 3, namely relativistic quantum mechanics or quantum field theory. This is based on the special relativistic theory of quantised fields, known to be very successful in situations where gravity can be neglected, though also subject to divergences. The theory has to be supplemented by an elaborate theoretical structure to deal with these divergences. Removal of divergences will be even more difficult in curved spacetimes.

A very powerful concept in quantum field theory is the idea of local or gauge symmetry, an expression of the fact that fundamental physical laws appear to be geometrical in nature. General relativity is a classic example, having the gauge symmetry associated with changing coordinates. The standard model of particle physics is also based upon gauge symmetries. Because of their importance, a whole chapter will be devoted to gauge symmetries, chapter 4.

Another feature of quantum field theory that has turned out to have many applications is quantum tunnelling through a potential barrier. Tunnelling is a purely quantum phenomenon with no classical analogue. Coleman (1977) was the first to show that when tunnelling occurs there generally exists a related solution to the field equations, modified by allowing time to be imaginary. The tunnelling rate N is related to the tunnelling solution by a simple formula

$$N = |A| \exp(-B). \tag{1.5}$$

The exponential factor B is the difference between the actions of the tunnelling solution and a solution where nothing happens. The factor A depends upon the quantum fluctuations about the tunnelling solution.

The existence of the tunnelling solution can be understood in terms of a gauge symmetry associated with freedom to choose a time coordinate. This symmetry can be introduced into any time independent system. By making the choice of time coordinate arbitrary, analytic continuation to imaginary time becomes possible, allowing tunnelling solutions.

The existence of quantum tunnelling also helps to explain why imaginary time is thought to be important in relation to gravity. Soon after the discovery that black holes radiate, Hartle and Hawking (1976) were able to show that the evaporation of particles from a black hole is a sort of tunnelling process in which matter tunnels from inside the horizon. It was this picture that originally suggested the usefulness of particle propagation on the Riemannian black hole spacetime. In a more speculative context, when quantising gravity, quantum fluctuations in the spacetime geometry should be able to tunnel into existence. These might conceivably be represented by imaginary-time solutions to the field equations known as gravitational instantons.

1.3　The inflationary universe

Other important questions of quantum gravity are how the universe began and how it developed its essential features. In 1981 Guth outlined a model that offers possible answers, which he called the inflationary universe. Problems with Guth's original model were resolved a year later (Albrecht and Steinhart 1982, Hawking and Moss 1982, Linde 1982). These models describe an era in which the potential energy of a scalar field ϕ (as yet undiscovered) drives the usual expansion of the universe into a rapid acceleration phase. This potential energy would be the source of all the matter and radiation in the universe today.

Without inflation, large-scale features of the universe, even scaled down to their size in the early universe, are still too large to be generated by the physical processes at work back then. The velocity of light limits the sphere of influence of any physical process at all compatible with notions of causality. In units where the velocity of light $c = 1$, regions of influence have sizes comparable to the timescale set by inverting the expansion rate H. The effect of inflation is that large-scale features of the universe scale down to sizes small enough to be influenced by microscopic physical phenomena such as quantum mechanics.

Discussions of quantum fluctuations of the scalar field in inflationary models are based closely on ideas developed from the study of quantum fluctuations around black holes. The correlation function for fluctuations in either the scalar field or the gravitational field on scales l, when $Hl \gg 1$, has the form

$$\xi(l) \sim \frac{\hbar H^2}{2\pi^2} \log Hl \qquad (1.6)$$

For comparison, the correlation function for massless fields in Minkowski spacetime falls off with distance as $\xi(l) \sim -l^{-2}$ at zero temperature, and approaches a constant $\xi(l) \sim \frac{1}{6}T^2$ at temperatures $T > 0$.

Shortly after the inflationary scenario was invented, it became clear that these quantum fluctuations are a possible source of the density fluctuations that later form galaxies and temperature fluctuations in the cosmic microwave background (Hawking 1982, Guth and Pi 1982, Bardeen et al. 1983). The amplitude for density perturbations can be related to the scalar field fluctuations by an extra factor, implying that the density perturbation amplitudes are set by the value of $H^2/\dot{\phi}$ during the inflationary era. This puts us in the position of being able to test the theory against observation. Even better, since both scalar perturbations and gravitational field fluctuations can be important for the cosmic microwave background fluctuations, these may present the first direct observational evidence of quantum gravitational phenomena.

Inflation has an even deeper connection with a quantum theory gravity. This is because inflation introduces the possibility that the observable universe was initially small enough to form in a quantum tunnelling event. According to this picture the universe tunnelled into existence from nothing. Such a process can be described using imaginary time. The tunnelling rate N described above might now be interpreted as a tunnelling probability for the whole universe. Different solutions to the gravitational field equations would allow us to compare the likelihood of different universes.

The challenge ahead lies in developing this possibility in more detail or inventing an alternative. One difficult issue is that the quantum description includes a superposition of many different possible universes. Somehow the different possibilities have to be partitioned into separate sets, and an explanation given of what the probabilities mean. This is the cosmological version of the problem of Schrödinger's cat. More details of this problem will be discussed in chapter 9.

If the quantum tunnelling picture can be justified, then the universe has no preferred beginning in time. The universe would be finite and self-contained. To reach a convincing statement of this sort we have to learn a great deal more about the nature of time and the meaning of quantum theory as well as developing a fundamental theory of the forces of Nature. We still have a long way to go.

2

Quantum theory and path integrals

The equivalence principle of general relativity states that the same laws of physics should be valid in the neighbourhood of any point in the universe. Attempts to combine quantum mechanics with general relativity face a number of problems. In the first place it is not possible to localise quantum phenomena to arbitrary small neighbourhoods of spacetime. In particular, the concept of particle number becomes vague on scales smaller than the Compton wavelength $\hbar/mc$ and we might therefore expect difficulties in defining particles when any radius of curvature of space is very small. Secondly, any quantum process will react back on the gravitational field through its energy. It will therefore be necessary, at some stage, to replace spacetime itself with a quantum description which takes fluctuations of the metric into account. This is one of the major problems for theoretical physics in our time.

There are many situations where the energy density and the Compton wavelengths are too small to have any effect in the region of interest. Most laboratory physics would be included in this regime. This is a good approximation too for the process of black hole evaporation, which will be discussed in a later chapter. For these situations a description of quantum fields in curved spacetimes has been developed, some of the basic ideas of which are presented in this and the next chapter.

2.1 Single-particle quantum mechanics

In elementary quantum mechanics, physical quantities correspond to operators acting on a Hilbert space of physical states of the system. Measurements are meant to deliver eigenvalues of these operators, with probabilities given by the squared modulus of an amplitude. These amplitudes will be denoted by

$$\langle \psi \mid \mathbf{A} \mid \psi \rangle. \tag{2.1}$$

The time development of states is linear and unitary. It can be represented by an operator $\mathbf{U}(t_1, t_0)$,

$$| \psi(t) \rangle = \mathbf{U}(t, t_0) \, | \, \psi(t_0) \rangle \qquad (2.2)$$

with the composition property

$$\mathbf{U}(t, t_0) = \mathbf{U}(t, t_1)\mathbf{U}(t_1, t_0). \qquad (2.3)$$

Taken together, these properties imply that the normalisation of the states is preserved. It also follows that

$$\mathbf{U}(t + \delta t, t) = 1 - \frac{i}{\hbar}\mathbf{H}\,\delta t, \qquad (2.4)$$

where $\mathbf{H}$ is a Hermitian operator. The differential form of time development can be obtained from

$$| \, \psi(t + \delta t) \rangle - | \, \psi(t) \rangle = \mathbf{U}(t + \delta t, t) \, | \, \psi(t) \rangle. \qquad (2.5)$$

After taking the limit $\delta t \to 0$,

$$i\hbar \frac{\partial}{\partial t} \, | \, \psi(t) \rangle = \mathbf{H} \, | \, \psi(t) \rangle. \qquad (2.6)$$

This is Schrödinger's equation, though the correspondence principle has to be invoked before $\mathbf{H}$ can be related to the classical Hamiltonian.

It may be one of those accidents of history that quantum theory began as a theory of operators acting on a Hilbert space of states. An equivalent formulation of quantum mechanics exists based upon transition amplitudes and path integrals. This approach is particularly direct and has proven to be a very powerful way of analysing quantum fields.

The path integral approach is based upon the observation that a classical particle follows a trajectory of stationary action under variation of the particle's path. In quantum theory the importance of the action is recognised by assigning an amplitude $e^{iS/\hbar}$ to each path. The total amplitude is obtained by summing over every path which is consistent with the conditions of a given experiment. In the limit that $\hbar \to 0$, the amplitude is dominated by the classical path with stationary action because of the cancellation of phases.

The amplitude to travel from position $\boldsymbol{x}$ at time t to position $\boldsymbol{x}'$ at time t' in terms of an integral over paths would be

$$\langle \boldsymbol{x}', t' | \boldsymbol{x}, t \rangle = \int_{t}^{t'} d\mu[\boldsymbol{x}]\mathcal{P}(\boldsymbol{x}', t' | \boldsymbol{x}, t)e^{iS[x]/\hbar}. \qquad (2.7)$$

The paths are labelled by $\boldsymbol{x}(\tau)$ with $t < \tau < t'$ and the distribution $\mathcal{P}(\boldsymbol{x}', t' | \boldsymbol{x}, t)$ fixes the endpoints to be at the chosen values, as shown in fig

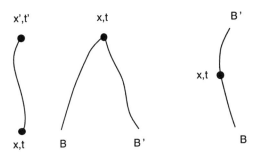

Figure 2.1 A selection of paths (a), (b) and (c) with time measured vertically.

2.1. The vertical bar in the expression $(x', t'|x, t)$ is meant to be read as 'and' in this context, without the connotation of products between state vectors.

Wavefunctions can also be introduced in this approach,

$$\langle x, t|\psi\rangle = \int_{-\infty}^{t} d\mu[x]\mathcal{P}(x, t)\psi(B)\, e^{iS[x]/\hbar} \tag{2.8}$$

and conjugate wavefunctions

$$\langle\psi|x, t\rangle = \int_{-\infty}^{t} d\mu[x]\mathcal{P}(x, t)\psi(B)^{*}\, e^{-iS[x]/\hbar}. \tag{2.9}$$

The state depends on a set of past boundary conditions B and varies with the choice of amplitudes assigned to B, as shown in figure 2.1.

Expectation values for measurements of some quantity $A(x, t)$ at a fixed time t would be

$$\langle\psi|\mathbf{A}|\psi\rangle = \int_{\Sigma} d\mu\mathcal{D}(x, t)\, A(x, t) \tag{2.10}$$

where $\mathcal{D}(x, t)$ is the propability distribution for the position coordinate $x \in \Sigma$. Using wavefunctions we would express this probability by the standard expression,

$$\mathcal{D}(x, t) = \langle\psi|x, t\rangle\langle x, t|\psi\rangle. \tag{2.11}$$

Using the path integral to rewrite the wavefunctions gives

$$\mathcal{D}(x,t) = \int_{-\infty}^{t} d\mu[x_1, x_2] \mathcal{P}_1(x,t) \mathcal{P}_2(x,t) \rho(B_1, B_2) e^{i(S[x_1]-S[x_2])/\hbar} \qquad (2.12)$$

where the past boundary conditions (B_1, B_2) are weighted by a function $\rho(B_1, B_2)$.

In general, we can set up a path integral over paths that are consistent with any set of conditions $\mathcal{C}$ which represent our preparation and observations of the system by inserting a corresponding projection function $\mathcal{P}[\mathcal{C}]$. The result is an amplitude $\langle \mathcal{C} \rangle$. The probability distribution would accordingly be generalised to a functional $\mathcal{D}[\mathcal{C}, \mathcal{C}']$. This is called the decoherence functional and has an important role to play in discussions of quantum theory. We shall meet it again in chapter 9.

With symmetry under time reversal, reversing the time in the conjugate amplitude and placing the boundary conditions B' in the future changes nothing but the sign of the action in the exponential. This allows for a simpler form of the expectation value than equation (2.10),

$$\langle \psi | \mathbf{A} | \psi \rangle = \int_{-\infty}^{\infty} d\mu[x] \rho(B, B') A(x(t), t) e^{iS[x]/\hbar} \qquad (2.13)$$

where the paths are taken from the past of t to the future. This is the form of path integral that is used most often in the theory of quantum fields.

2.2   Evaluating the path integral

For elementary quantum mechanics, the path integral can be defined by a simple limiting procedure, splitting the time interval $[-\infty, t]$ into steps of size $\delta\tau$ and splitting the path into line segments (figure 5.3). The simplest way to integrate over each path is to integrate over the ends of the segments. The integration measure for $\langle x, t | \psi \rangle$ is then

$$\int_{-\infty}^{t} d\mu[x] = \prod_{\tau=-\infty}^{t} \int \mu_R \, dx_\tau \qquad (2.14)$$

where μ_R is a normalisation factor. A distribution $\mathcal{P}(x,t) = \delta(x_t - x)$ fixes the endpoint.

If the action depends only on the position and its first time derivative then it can also be split into segments,

$$S = \sum_{\tau=-\infty}^{t-\delta\tau} \widehat{S}(x_\tau, x_{\tau+\delta\tau}) \qquad (2.15)$$

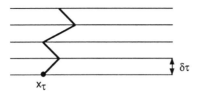

Figure 2.2 Approximating a continuous world–line by line segments.

by replacing time derivatives with finite differences.

The measure has an important separation property where an intermediate time t' is singled out,

$$\int_{-\infty}^{t} d\mu[x] = \mu_R \int dx' \int_{t'}^{t} d\mu[x]\mathcal{P}(x',t') \int_{-\infty}^{t'} d\mu[x]\mathcal{P}(x',t'). \qquad (2.16)$$

The exponential integrand also factorises,

$$\exp\left(\frac{i}{\hbar}\sum_{\tau=-\infty}^{t-\delta\tau}\widehat{S}\right) = \exp\left(\frac{i}{\hbar}\sum_{\tau=t'}^{t-\delta\tau}\widehat{S}\right)\exp\left(\frac{i}{\hbar}\sum_{\tau=-\infty}^{t'-\delta\tau}\widehat{S}\right) \qquad (2.17)$$

making it possible to decompose the entire path integral, or equivalently

$$\langle x,t|\psi\rangle = \mu_R \int dx' \langle x,t|x',t'\rangle\langle x',t'|\psi\rangle. \qquad (2.18)$$

The relationship between the path integral and the time evolution operators can be set up by introducing a Hilbert space with an orthogonal basis $|x\rangle$. We define

$$\mathbf{U}(t,t_0) = |x\rangle\langle x,t|x_0,t_0\rangle\langle x_0| \qquad (2.19)$$

The composition property follows from equation (2.18). It follows that the time evolution will be consistent with the basic axioms of the operator approach. The following example shows how the Hamiltonian operator follows from the form of the action.

Example: The path integral for a particle in one dimension

The action for a particle in one dimension interacting with a potential $V(x)$ is,

$$S = \int_{-\infty}^{t} (\tfrac{1}{2}m\dot{x}^2 - V(x))d\tau. \qquad (2.20)$$

After splitting the paths into segments the action for each line segment would be

$$\widehat{S}(x_\tau, x_{\tau+\delta\tau}) = \left(\frac{m}{2} \left(\frac{x_{\tau+\delta\tau} - x_\tau}{\delta\tau} \right)^2 - V \left(\frac{x_{\tau+\delta\tau} + x_\tau}{2} \right) \right) \delta\tau. \quad (2.21)$$

Consider in particular what happens to the wavefunction for an infinitesimal time step beyond t. The measure decomposes into

$$\int_{-\infty}^{t+\delta t} d\mu[x] \mathcal{P}(x, t+\delta t) = \mu_R \int_{-\infty}^{\infty} dx_t \int_{-\infty}^{t} d\mu[x] \mathcal{P}(x_t, t). \quad (2.22)$$

By separating off the last term in the action we can write

$$\langle x, t+\delta t \mid \psi \rangle = \mu_R \int_{-\infty}^{\infty} dx_t \, e^{i\widehat{S}(x_t, x)/\hbar} \langle x_t, t \mid \psi \rangle. \quad (2.23)$$

Setting $\xi = x_t - x$ and using equation (2.21) gives

$$\langle x, t+\delta t | \psi \rangle = \mu_R \int_{-\infty}^{\infty} d\xi \, e^{im\xi^2/(2\hbar\delta t)} e^{-iV(x)\delta t/\hbar} \langle x+\xi, t | \psi \rangle. \quad (2.24)$$

It is now possible to expand the amplitudes in powers of δt and ξ. In the limit that $\delta t \to 0$ and $\xi^2 = O(\delta t)$, we perform the integrals and find that $\mu_R \to (2\pi/m\hbar\delta t)^{1/2}$ from the leading-order term. The next order gives

$$i\hbar \frac{\partial}{\partial t} \langle x, t | \phi \rangle = \left(-\frac{\hbar^2}{2m} \frac{\partial^2}{\partial x^2} + V(x) \right) \langle x, t | \phi \rangle \quad (2.25)$$

demonstrating that the path integral is equivalent to the Schrödinger equation.

2.3 Relativistic quantum theory

The first step towards a Lorentz or generally covariant quantum theory is to replace the action and the Hamiltonian by covariant expressions. It is natural to work in spacetime, where the location of a particle is described by four coordinates $x^a = (t, \boldsymbol{x})$.

The relativistic version of Schrödinger's equation is the wave equation

$$\left(-\hbar^2 \nabla^2 + m^2 \right) \psi = 0 \quad (2.26)$$

where $\nabla^2 = g^{ab} \nabla_a \nabla_b$ and ∇_a is the connection of the spacetime metric g_{ab}.

For the moment, consider what happens when spacetime is flat. There are plane-wave solutions to the wave equation of the form

$$\exp(i\boldsymbol{p} \cdot \boldsymbol{x} - iEt)/\hbar \quad (2.27)$$

where the energy E satisfies $E^2 = p^2 + m^2$. This surface in momentum space is sometimes called the mass shell. In the classical theory of fields, negative-as well as positive-frequency solutions are needed to make up a complete set of solutions to the wave equation. In quantum theory the negative-energy solutions are a potential embarrassment. The reason is that quantum states can decay into lower energy states under the influence of interactions. It is essential therefore that there should be a lowest-energy state or the decays will never stop.

The propagator $G(\mathbf{x}, \mathbf{x}') = \langle \mathbf{x}' \mid \mathbf{x} \rangle$ evolves particle states forwards in time and must be constructed in a way that prevents the spread of these negative-energy waves. It satisfies the equation for Green functions,

$$\left(-\hbar^2 \nabla^2 + m^2\right) G(\mathbf{x}, \mathbf{x}') = \hbar^2 \delta(\mathbf{x}, \mathbf{x}'). \tag{2.28}$$

Fourier transforms can be used to solve the equation,

$$G(\mathbf{x}, \mathbf{x}') = \frac{1}{\hbar^2} \int_C \frac{d^4 p}{(2\pi)^4} \, G(\mathbf{p}) e^{i \mathbf{p} \cdot (\mathbf{x} - \mathbf{x}')/\hbar} \tag{2.29}$$

implies $(p^2 + m^2)G(\mathbf{p}) = 1$. The general solution to this is

$$G(p) = 1/(p^2 + m^2) + f(p)\delta(p^2 + m^2) \tag{2.30}$$

with some freedom involving the definition of $f(p)$.

In practice this ambiguity can be transformed into a variety of integration contours in the complex E plane, as shown in figure 2.3. If $t' < t$, the exponential allows closure of the contour of integration in the lower half-plane. The retarded propagator G_R has no poles inside this contour and therefore

$$G_R(x, x') = 0 \quad \text{for } t' < t. \tag{2.31}$$

This propagator is used uniformly in classical field theory where it is necessary to ensure that cause should always precede effect.

The most useful propagator in quantum field theory is the Feynman propagator because it has only positive frequencies (in t') for $t' > t$ and negative frequencies when $t' < t$. The particle wavefunctions themselves are not measurable in any simple way and problems with causality do not seem to arise (as far as we know).

The Feynman propagator is often combined with analytic continuation to imaginary time, $t = -i\tau$. The new propagator is called the Wick-rotated or Euclidean propagator,

$$G_F(\mathbf{x}, t, \mathbf{x}', t') = iG_E(\mathbf{x}, it, \mathbf{x}', it'). \tag{2.32}$$

The Euclidean propagator satisfies an elliptic equation, and it is often convenient to use this analytic continuation of t and E for calculations in

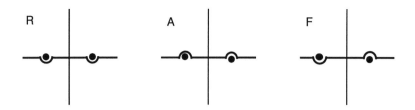

Figure 2.3   Integration contours in the complex E plane.

flat spacetime. The analogous procedure in curved spaces or spaces with non-trivial topology has physical consequences and shall be considered later in connection with finite-temperature field theory and quantum gravity. (The same notation G_E will be used for the propagator on these Riemannian spaces.)

Although the Feynman propagator can be obtained from a path integral (after analytic continuation), there is no operator analogue acting on a Hilbert space of *single*-particle states. This can be associated with the existence of pair-creation and annihilation, shown as world-lines in figure 2.4. Also shown is a rendering of the path as a single-particle world-line that propagates in both the future and past directions of time. Failure of the composition rule for the time evolution operator $\mathbf{U}(t, t_0)$ follows because the path crosses some surfaces of constant time more than once.

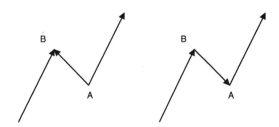

Figure 2.4   Particle world-line for pair creation followed by annihilation.

In flat space the resolution of this problem is to introduce many-particle states,

$$| \{n_i \text{ particles with wavefunction } f_i\}\rangle \qquad (2.33)$$

where the f_i are positive-frequency solutions to the wave equation. The number of particles in a state can be raised with a creation operator a_i^+, where

$$[\mathbf{a}_i, \mathbf{a}_j] = 0, \quad [\mathbf{a}_i, \mathbf{a}_j^+] = \delta_{ij} \qquad \text{for bosons} \qquad (2.34)$$

$$\{\mathbf{a}_i, \mathbf{a}_j\} = 0, \quad \{\mathbf{a}_i, \mathbf{a}_j^+\} = \delta_{ij} \qquad \text{for fermions}. \qquad (2.35)$$

(The anticommuting fermionic operators imply that $\mid f_i, f_j \rangle = - \mid f_j, f_i \rangle$.) These states form a basis for the Hilbert space of the system. A general state contains a large amount of information about the relative phases of components in different basis states, and only after particular measurements might it be possible to talk of reduction to the simpler concepts of 'a particle' or 'a field'.

Time evolution is described by the Feynman propagator, so that

$$\langle f_j(\mathbf{x}')|f_i(\mathbf{x})\rangle = \int d\mu(\boldsymbol{x}, \boldsymbol{x}') f_i(\mathbf{x}) G_F(\mathbf{x}, \mathbf{x}') f_j(\mathbf{x}'). \qquad (2.36)$$

This amplitude vanishes if $t' < t$ (as seen by completing the contour in the contour integral representation).

2.4   Fermions

The characteristic property of fermion fields is that they change sign under a rotation through 2π. It is therefore necessary to use a larger group than the Lorentz group to describe relativistic fermion fields. This group is Spin(3,1).

Spin(3,1) is defined in terms of gamma matrices γ^a. These are a set of four 4×4 complex matrices which satisfy

$$\{\gamma^a, \gamma^b\} = -2g^{ab} \quad \text{and} \quad \gamma^{a\dagger} = \gamma^0 \gamma^a \gamma^0. \qquad (2.37)$$

A matrix $\mathbf{S}$ belongs to Spin(3,1) if

$$\mathbf{S}^{-1}\gamma^a\mathbf{S} = \Lambda^a{}_b \gamma^b \qquad (2.38)$$

for some Λ in the Lorentz group. Spin(3,1) is a double covering of the Lorentz group because $\mathbf{S}$ and $-\mathbf{S}$ correspond to the same Lorentz transformation.

There are many representations of the γ matrix algebra. The chiral representation with respect to an orthonormal basis e_a is

$$\gamma^0 = \begin{pmatrix} 0 & 1 \\ 1 & 0 \end{pmatrix} \qquad \gamma^i = \begin{pmatrix} 0 & \sigma^i \\ -\sigma^i & 0 \end{pmatrix} \qquad (2.39)$$

where σ^i are the Pauli matrices.

Electrons and many other fermionic particles are represented by Dirac spinors belonging to the space on which the defining representation of

Spin(3,1) acts. The four components correspond to the spin-up and spin-down states of the fermion and its antiparticle partner. The existence of γ matrices makes it possible to construct a first-order covariant equation for fermion fields, which is the famous Dirac equation,

$$(i\hbar\gamma^a\nabla_a - m)\,\psi = 0. \tag{2.40}$$

In curved space, the covariant derivative contains a connection term to take account of Spin(3,1) transformations. In general connection terms belong to the Lie algebra of the group that acts on the fields. This will be explained more fully in chapter 4. The Lie algebra generators of Spin(3,1) are $\Sigma_{ab} = \frac{1}{8}i[\gamma_a, \gamma_b]$, that is $S = I + i\epsilon^{ab}\Sigma_{ab}$ is a member of Spin(3,1) for small matrices ϵ. This can be verified by direct substitution into equation (2.38) and gives an introduction to the kind of matrix manipulations which are very common in spinor calculations.

The spin group should be related to O(3,1) rotations of the local Lorentz frame by identifying their respective Lie algebra coefficients. The connection coefficients for the O(3,1) rotations are the ordinary spacetime connection coefficients measured in an orthonormal basis e_a. The covariant derivative is therefore

$$\nabla_a = e_a{}^\mu\left(\partial_\mu - i\omega^{cd}{}_\mu\Sigma_{cd}\right). \tag{2.41}$$

The spacetime connection coefficients $\omega^{cd}{}_\mu$ are described in appendix B.

The Dirac equation can also be converted into a second-order equation using

$$(-i\hbar\gamma\cdot\nabla - m)(i\hbar\gamma\cdot\nabla - m) = -\hbar^2\nabla^2 + \tfrac{1}{4}\hbar^2 R + m^2. \tag{2.42}$$

In this form the equation resembles the scalar wave equation except that the spin connection must be included in the covariant derivative.

In flat space, there are two positive-energy plane-wave solutions to the Dirac equation and two negative-energy solutions. The propagator $S(\mathbf{x}, \mathbf{x}')$ can be expressed as before in terms of the momentum space propagator $S(p)$. We can reduce the amount of work that we have to do by writing the Dirac propagator in terms of the scalar propagator, $S(\mathbf{p}) = (\gamma\cdot\mathbf{p} - m)G(\mathbf{p})$. Advanced, retarded and Feynman propagators can all be obtained using the appropriate integration contour in the complex energy plane.

Complex conjugation of Dirac spinor space produces an equivalent representation of Spin(3,1). That is to say there is a matrix C for which $S^* = CSC^{-1}$, or $\gamma^{a*} = C\gamma^a C^{-1}$, and this matrix is called the charge conjugation matrix. Another form of conjugation is important and arises due to the fact that the matrices S are not unitary, but satisfy $S^\dagger\gamma^0 = \gamma^0 S^{-1}$. Because of this property it is useful to define the Dirac conjugate $\bar{\psi} = \psi^\dagger\gamma^0$, and then $\bar{\psi}\psi$ is Spin(3,1) invariant.

The double covering of the proper Lorentz group SO(3,1), the group which excludes parity transformations $\mathbf{x} \to -\mathbf{x}$, is the special linear group

SL(2,C). The space on which this group acts is called Weyl spinor space and represents particles such as neutrinos whose behaviour is not symmetric under parity transformations. The reduction of a representation of Spin(3,1) to SL(2,C) is accomplished by chiral projection matrices $P_L = \frac{1}{2}(\mathbf{I} + \gamma_5)$, where $\gamma_5 = i\gamma^0\gamma^1\gamma^2\gamma^3$. This is simple to see in the chiral γ matrix representation, where the SL(2,C) subgroup of Spin(3,1) is block diagonal,

$$\mathbf{S} = \begin{pmatrix} S_L & 0 \\ 0 & S_L^\dagger \end{pmatrix} \quad \psi = \begin{pmatrix} \lambda^A \\ \eta_{A'} \end{pmatrix} \quad \mathbf{C} = \begin{pmatrix} \epsilon_{AB} & 0 \\ 0 & \epsilon^{A'B'} \end{pmatrix}. \tag{2.43}$$

The tensors ϵ_{AB} and $\epsilon^{A'B'}$ are antisymmetric with $\epsilon_{12} = 1$.

The theory of fermion fields in Riemannian space, with metric signature $(++++)$, is important if we are to understand the analytic continuation of the Feynman propagator. Riemannian Dirac spinors are associated with the group Spin(4) which is a double covering of the orthogonal group O(4). Riemannian γ matrices have a Hermitian representation

$$\gamma^0 = \begin{pmatrix} 0 & 1 \\ 1 & 0 \end{pmatrix} \quad \gamma^i = \begin{pmatrix} 0 & i\sigma^i \\ -i\sigma^i & 0 \end{pmatrix}. \tag{2.44}$$

Consequently, the Riemannian propagator in flat space satisfies

$$(i\gamma \cdot \mathbf{p} - m)S_E(\mathbf{p}) = 1. \tag{2.45}$$

The double covering of the special orthogonal group SO(4) is the group SU(2) × SU(2). Representations of Spin(4) are reduced by the chiral projection $P_L = \frac{1}{2}(\mathbf{I} + \gamma_5)$, where now $\gamma_5 = \gamma^0\gamma^1\gamma^2\gamma^3$. This can be seen with the chiral representation, where

$$\mathbf{S} = \begin{pmatrix} S_L & 0 \\ 0 & S_R \end{pmatrix} \quad \psi = \begin{pmatrix} \lambda^A \\ \eta_{A'} \end{pmatrix} \quad \mathbf{C} = \begin{pmatrix} \epsilon_{AB} & 0 \\ 0 & \epsilon^{A'B'} \end{pmatrix}. \tag{2.46}$$

The components S_L and S_R are independent SU(2) matrices.

In the Riemannian case charge conjugation requires $\gamma^{a*} = -\mathbf{C}\gamma^a\mathbf{C}^{-1}$. The representation of Spin(4) associated with the Hermitian γ matrices is unitary and $\psi^\dagger\psi$ is invariant. Dirac conjugation can now be identified with Hermitian conjugation, $\overline{\psi} = \psi^\dagger$.

2.5 Relativistic path integration

The covariant path integral for a relativistic particle should sum over paths $\mathbf{x}(\tau)$ of the particle through a spacetime manifold with metric $\mathbf{g}$. In the Lorentzian case paths like those shown in figure 2.4 can be omitted by restricting the sum to time-like paths. In this case the action of a freely moving particle is proportional to the proper time along the world-line,

$$S = -m \int_0^1 \left(-g_{ab}\dot{x}^a \dot{x}^b\right)^{1/2} d\tau. \tag{2.47}$$

The details given below show that the path integral will give a retarded propagator.

Example: Direct evaluation of the relativistic path integral

The path integral is defined by splitting the path into segments of parameter length $\delta\tau$,

$$G(\mathbf{x}_0, \mathbf{x}_1) = \int_0^1 d\mu[\mathbf{x}]\mathcal{P}(\mathbf{x}_1, 1|\mathbf{x}_0, 0)\, e^{iS[\mathbf{X}]/\hbar}. \tag{2.48}$$

If the action is to be real, the segments must be time-like and $\mathbf{x}_1$ must be confined to the causal future (or past) of $\mathbf{x}_0$. The measure

$$d\mu[\mathbf{x}]\mathcal{P}(\mathbf{x}_1, 1|\mathbf{x}_0, 0) = \prod_{\tau=\delta\tau}^{1-\delta\tau} (-\det g_{ab})^{1/2} dx_\tau. \tag{2.49}$$

The action can also be split up,

$$S[\mathbf{x}(\tau)] = \sum_{\tau=0}^{1-\delta\tau} S\left(\mathbf{x}_\tau, \mathbf{x}_{\tau+\delta\tau}\right) \tag{2.50}$$

replacing $\dot{\mathbf{x}}$ by a finite difference. In general, particularly when g_{ab} is not a constant, the form of this decomposition is not unique. It is important, at the very least, to ensure that the decomposition is independent of coordinates.

The way in which to ensure coordinate independence is to define S to be the action of a classical path between the endpoints of the segment. Then results from classical dynamics tell us that the momentum at the endpoint $\mathbf{x}_{\tau+\delta\tau}$ of the segment is

$$p_a(\mathbf{x}_\tau, \mathbf{x}_{\tau+\delta\tau}) = \frac{\partial S(\mathbf{x}_\tau, \mathbf{x}_{\tau+\delta\tau})}{\partial x_{\tau+\delta\tau}^a} \tag{2.51}$$

and also that

$$g^{ab}(\mathbf{x}_{\tau+\delta\tau})p_a(\mathbf{x}_\tau, \mathbf{x}_{\tau+\delta\tau})p_b(\mathbf{x}_\tau, \mathbf{x}_{\tau+\delta\tau}) + m^2 = 0. \tag{2.52}$$

The first of these equations allows us to differentiate $G(\mathbf{x}_0, \mathbf{x}_1)$,

$$-i\hbar\frac{\partial}{\partial x_1^a} G(x_0, x_1) = \int_0^1 d\mu[x]\, p_a(x_{1-\delta\tau}, x_1)e^{iS[x]/\hbar}. \tag{2.53}$$

It is also possible to differentiate the momentum by relating S to the geodesic interval σ between the endpoints of the segment,

$$\sigma = \int_\tau^{\tau+\delta\tau} \left(-g_{ab}\dot{x}^a\dot{x}^b\right)^{1/2} d\tau \qquad (2.54)$$

and therefore $S = -m\sigma$. It follows that

$$\nabla^a p_a(x_\tau, x) = -m\nabla^2\sigma = O(\sigma) \qquad (2.55)$$

as $\sigma \to 0$. Covariant derivatives are crucial here because in general $\sigma_{,ab} = O(\sigma^{-1})$ and the argument below would fail.

Parenthetically, we note that for small σ it is possible to write

$$S(\mathbf{x}_\tau, \mathbf{x}_{\tau+\delta\tau}) = -m \left(-g_{ab}(\mathbf{x}_\tau)(x_{\tau+\delta\tau}^a - x_\tau^a)(x_{\tau+\delta\tau}^b - x_\tau^b)\right)^{1/2} + O(\sigma^2) \qquad (2.56)$$

and then

$$p_a(\mathbf{x}_\tau, \mathbf{x}_{\tau+\delta\tau}) = m g_{ab}(\mathbf{x}_\tau)(x_{\tau+\delta\tau}^b - x_\tau^b)/\sigma + O(\sigma). \qquad (2.57)$$

This shows that the leading term for p_a is still the finite difference, which we expect. The $O(\sigma)$ terms are fixed by our particular choice of S.

Returning to the argument, it is now a simple matter to derive the relativistic wave equation. Differentiate $G(\mathbf{x}_0, \mathbf{x}_1)$ one more time, writing $\mathbf{p}(\mathbf{x}_1) = \mathbf{p}(\mathbf{x}_{1-\delta\tau}, \mathbf{x}_1)$

$$-\hbar^2 g^{ab}(\mathbf{x}_1)\nabla_a\nabla_b G(\mathbf{x}_0, \mathbf{x}_1) = \int_0^1 d\mu[x] \, g^{ab}(\mathbf{x}_1) p_a(\mathbf{x}_1) p_b(\mathbf{x}_1) e^{iS[x]/\hbar}. \qquad (2.58)$$

From the Hamilton–Jacobi equation (2.52),

$$(-\hbar^2\nabla^2 + m^2)G(\mathbf{x}_0, \mathbf{x}_1) = 0 \qquad (2.59)$$

when t' is not equal to t. The condition that $G(\mathbf{x}_0, \mathbf{x}_1)$ is non-zero only when x_1 is to the future of x_0 implies that $G(\mathbf{x}_0, \mathbf{x}_1)$ is the retarded Green function. We can obtain the Euclidean propagator, and therefore the Feynman Green function by analytic continuation, starting from a Riemannian spacetime manifold.

An important consideration is the way of replacing the continuous action integral by a discrete sum. Basing the discretisation on the action of a classical line segment leads to a covariant wave equation. However, this geometrical approach can still give wave equations that differ by terms depending on the curvature of g_{ab}. This ambiguity can arise because it is possible to add a term $\frac{1}{2}\zeta\hbar R\sigma^2$, where ζ is a constant, to the action. This does not affect the classical limit as it vanishes when $\hbar \to 0$ and it is certainly covariant. The wave equation becomes

$$(-\hbar^2\nabla^2 + \zeta\hbar R + m^2)G(x_0, x_1) = 0. \qquad (2.60)$$

The ambiguity is more evident still in operator approaches to the wave equation. In chapter 4 the following classical Hamiltonian is given,

$$H = \tfrac{1}{2}(g^{ab}p_a p_b + m^2).$$ (2.61)

In quantum theory the momentum operator $p_a = -i\hbar\partial/\partial x^a$. Therefore the form of the Hamiltonian operator depends on the ordering of p_a and g_{ab} whenever the metric is not constant. This ambiguity can be present even in flat space if the coordinate system is curvilinear, or if there are constraints on the variables that restrict the system to a surface. This last point implies that factor ordering is an issue even for non-relativistic quantum systems.

Factor ordering can be chosen in a way that imposes covariance of the equations under coordinate changes. One particular example would be

$$H = \tfrac{1}{2}(g^{-1/2}p_a g^{1/2}g^{ab}p_b + m^2).$$ (2.62)

This is equivalent to

$$H = \tfrac{1}{2}(-\hbar^2\nabla^2 + m^2).$$ (2.63)

There is the same freedom to add extra curvature terms that vanish in the classical limit that we saw above.

3

Quantum field theory

The historical debate on the nature of light ended with the discovery of quantum field theory, because the quantum theory of relativistic particles and the quantum theory of fields turned out to be identical. In quantum field theory the basis of states can be wave-like or particle-like. States in general are superpositions, capable of representing a variable number of particles of every possible momentum, with all of the additional phase information that can exist between these fixed particle-number states.

This chapter will be concerned with quantising matter fields when the spacetime geometry is fixed and independent of time. The aim is to be able to calculate the behaviour of field expectation values and find out how they might begin to react back on the geometry. Many of the techniques explained here were developed by DeWitt in the 1960s, but we will follow Dowker (Dowker and Critchley 1976) in using a generalised ζ-function to remove divergences.

3.1 Classical field theory

Classical field theory is governed by relativistic wave equations. These equations can be derived from an action principle. In each case the action S is the volume integral of a Lagrangian function depending on the field and its first space-time derivative. The action is stationary under variations of the field when the field satisfies the field equation. A list of Lagrangians is given in table 3.1.

Different types of field are classified by their spin. The photon, for example, is a spin-1 field and can be represented by the four-vector potential $\mathbf{A}$. The photon Lagrangian is given in terms of the field strength tensor $\mathbf{F}$. This Lagrangian will be studied in the next chapter.

Before considering variation of the action it is worth considering integration by parts. This follows from a generalised form of the divergence theorem,

$$\int_{\mathcal{M}} d\mu \, \boldsymbol{\nabla} \cdot \mathbf{X} = \int_{\partial \mathcal{M}} d\mu \, \mathbf{X} \cdot \mathbf{n}. \tag{3.1}$$

Table 3.1   Wave equations and Lagrangians for various free fields.

Field	Wave equation	Lagrangian
scalar	$-\nabla^2 + m^2$	$-\frac{1}{2}\boldsymbol{\nabla}\phi \cdot \boldsymbol{\nabla}\phi - \frac{1}{2}m^2\phi^2$
spinor	$i\boldsymbol{\gamma} \cdot \boldsymbol{\nabla} - m$	$\overline{\psi}(i\boldsymbol{\gamma} \cdot \boldsymbol{\nabla} - m)\psi$
vector	$-\delta_a{}^b\nabla_c\nabla^c + \nabla_a\nabla^b + R_a{}^b$	$-\frac{1}{4}F_{ab}F^{ab}$

Replacing $\mathbf{X}$ by a product of tensors $\mathbf{uv}$ gives,

$$\int_{\mathcal{M}} d\mu\,(\mathbf{u} \cdot \boldsymbol{\nabla})\,\mathbf{v} = -\int_{\mathcal{M}} d\mu\,(\boldsymbol{\nabla} \cdot \mathbf{u})\,\mathbf{v} + \int_{\partial\mathcal{M}} d\mu\,(\mathbf{u} \cdot \mathbf{n})\,\mathbf{v}. \qquad (3.2)$$

Variation of the scalar field action with $\delta\phi = 0$ on the boundary gives

$$\delta S = \int_{\mathcal{M}} d\mu \left\{ \frac{\partial\mathcal{L}}{\partial\phi}\delta\phi + \frac{\partial\mathcal{L}}{\partial\phi_{;a}}\delta\phi_{;a} \right\}. \qquad (3.3)$$

Integration by parts allows this to be written in terms of an integration kernel, called the functional derivative,

$$\delta S = \int_{\mathcal{M}} d\mu\, \frac{\delta S}{\delta\phi}\delta\phi. \qquad (3.4)$$

In this case,

$$\frac{\delta S}{\delta\phi} = \frac{\partial\mathcal{L}}{\partial\phi} - \left(\frac{\partial\mathcal{L}}{\partial\phi_{;a}}\right)_{;a}. \qquad (3.5)$$

The action is stationary when this expression vanishes. The functional derivative satisfies Leibnitz' rule and, with volume integrals, the chain rule for derivatives.

Real scalar fields are the simplest kind of field. They form an important constituent in modern theories of fundamental physics, although there is still an absence of any direct evidence for their existence. Fields are described as either free, with linear wave equations, or interacting. The rest of this chapter will be devoted to the quantum theory of such fields.

3.2   Quantum field theory with imaginary time

In the special case where the spacetime is stationary there is a conserved energy defined on any spatial hypersurface Σ, with normal $\boldsymbol{\omega}$,

$$E = \int_{\Sigma} d\mu\,\boldsymbol{\omega}(\mathbf{T}) \qquad (3.6)$$

where $\mathbf{T}$ is the stress–energy tensor. A quantum state of minimum energy would naturally be identified as the ground state.

Time translation symmetry would also make it possible to use analytic continuation of the time coordinate to form a related Riemannian space (i.e. positive definite metric). Propagators on the Riemannian space are often simpler to use than propagators on the real spacetime. In flat spacetime the Euclidean or Riemannian propagator is the analytic continuation of the Feynman propagator.

If the propagation of particles can be recovered from a Riemannian propagator then the whole quantum theory should be capable of being expressed in terms of the Riemannian space. Continuation of the field-theory action produces a Riemannian or Euclidean action I,

$$iS = \int \mathcal{L} \sqrt{g} d^3x d(it) = -I. \tag{3.7}$$

Vacuum expectation values should be obtained by a path integral,

$$\langle 0 \mid \mathbf{A} \mid 0 \rangle = \frac{1}{Z} \int d\mu[\phi] \mathbf{A} e^{-I} \tag{3.8}$$

where Z is a normalisation factor. The integral extends over all field configurations on the fixed spacetime background. Boundary conditions can be imposed on the fields in the path integral which correspond to the boundary conditions on the propagator. The fields in the path integral may approach their minimum energy values at infinity, or the Riemannian space may alternatively have surfaces on which a more general class of boundary conditions on the fields have to be specified.

In many cases, such as with the black hole spacetime, the situation is complicated by a change in the global structure of the space when going from Lorentzian to Riemannian sections. If it is not possible to identify the initial or final surfaces in the Riemannian section then it becomes difficult in these cases to define transition amplitudes by a Riemannian path integral. The path integral may still define an expectation value of some kind. In the black hole case we shall see in a later chapter that the expectation value can be associated with a thermal partition function.

3.3   Vacuum amplitudes

In flat space we take it for granted that the vacuum has zero energy and pressure. In curved space this need no longer be the case. Properties of the vacuum can be found by introducing the vacuum amplitude,

$$Z[\mathbf{g}] = \int d\mu[\phi] e^{-I[g,\phi]/\hbar} \tag{3.9}$$

defined for fields on a fixed gravitational background with metric **g**. We shall see how the vacuum amplitude is used to find the expectation value of the stress–energy tensor T_{ab}.

Information can be extracted from the vacuum amplitude by using functional calculus. Variation of the background metric defines the functional derivative $\delta Z/\delta \mathbf{g}$,

$$\delta Z = \int_{\mathcal{M}} d\mu \frac{\delta Z}{\delta g_{ab}} \delta g_{ab}. \tag{3.10}$$

In many situations we also assume that it is possible to differentiate through the path integral. If this is the case then

$$\frac{\delta Z}{\delta g_{ab}} = -\frac{1}{\hbar} \int d\mu[\phi] \frac{\delta I}{\delta g_{ab}} e^{-I[g,\phi]/\hbar}. \tag{3.11}$$

The variation of the classical action with respect to the metric defines the stress–energy tensor,

$$T^{ab} = -2 \frac{\delta I}{\delta g_{ab}}. \tag{3.12}$$

With the normalisation constant,

$$\langle T^{ab} \rangle = 2\hbar \frac{1}{Z} \frac{\delta Z}{\delta g_{ab}}. \tag{3.13}$$

This is usually rewritten,

$$\langle T^{ab} \rangle = -2 \frac{\delta \Gamma}{\delta g_{ab}} \tag{3.14}$$

using the logarithm of the vacuum amplitude, $\Gamma[\mathbf{g}] = -\hbar \log Z[\mathbf{g}]$.

Gravity has been regarded so far as a classical field. All of the fields representing the other forces of Nature are known to be quantised. The gravitational field could be quantised by introducing a larger path integral which also included a sum over metrics,

$$Z = \int d\mu[g] e^{-(I_g[g]+\Gamma[g])/\hbar}. \tag{3.15}$$

If we can ignore the gravity fluctuations then the path integral would be dominated by stationary points of the exponent (3.15),

$$\frac{\delta I_g}{g_{ab}} + \frac{\delta \Gamma}{g_{ab}} = 0. \tag{3.16}$$

The variation of the Einstein action produces the Einstein tensor,

$$G^{ab} = 16\pi G \frac{\delta I_g}{\delta g_{ab}}. \tag{3.17}$$

This implies that

$$G^{ab} = 8\pi G\langle T^{ab}\rangle. \tag{3.18}$$

Therefore the effective action $\Gamma[g]$ can be used to determine self-consistent backgrounds including back-reaction of quantised fields on the curvature of spacetime. (This effective action is a simplified form of the quantity usually called the effective action, which will be introduced later.)

Evaluation of Z in simple examples is based upon the formula for the Gaussian integral,

$$\int_{R^D} d\mu \, e^{-\mathbf{x}\cdot\mathbf{A}\cdot\mathbf{x}} = \pi^{D/2} \left(\det \mathbf{A}\right)^{-1/2} \tag{3.19}$$

where $\mathbf{A}$ is a matrix. The problem consists of generalising this formula to infinite-dimensional operators.

Consider a free scalar field with field equation $\Delta\phi = 0$ on a compact manifold $\mathcal{M}$. The field equation can be generated by variation of the action,

$$I = \tfrac{1}{2} \int_{\mathcal{M}} d\mu \, \phi\Delta\phi \tag{3.20}$$

provided that ϕ is held fixed on any boundaries.

Now consider a path integral over fields that can be expanded in terms of normalised eigenfunctions u_n of Δ with the same boundary conditions. The action can be written as a function of the expansion coefficients c_n and the eigenvalues λ_n,

$$I = \tfrac{1}{2} \sum_{n=1}^{\infty} c_n^2\lambda_n. \tag{3.21}$$

In the previous chapter the measure was defined using discrete time steps. In this case too we could replace the manifold by a lattice of N points x_n. The difference operator corresponding to Δ would have N eigenfunctions u_n, with

$$\int \prod_{n=1}^{N} (\mu_R d\phi_n) = \int \prod_{n=1}^{N} (\mu_R dc_n) \tag{3.22}$$

due to orthogonality. Therefore

$$Z[g] = \int \prod_{n=1}^{\infty} (\mu_R dc_n) \exp\left(-\tfrac{1}{2} \sum_{n=1}^{\infty} c_n^2\lambda_n\right) \tag{3.23}$$

where μ_R is a normalisation factor. Performing each of the separate integrals gives

$$Z[g] = \prod_{n=1}^{\infty} \left(\frac{\lambda_n}{\mu_R^2}\right)^{-1/2}. \tag{3.24}$$

The problem with this expression is that $\lambda_n \sim \sqrt{n}$ for large n. Some kind of truncation procedure is needed to obtain a sensible result. This is known as regularisation.

It will be useful to define

$$\det \Delta = \prod_n \lambda_n \qquad (3.25)$$

in some way, then

$$\Gamma = \tfrac{1}{2}\hbar \log \det \Delta. \qquad (3.26)$$

We shall see how this can be done using a generalised version of the Riemann ζ-function.

3.4   Zeta-function regularisation

A great deal can be learned about the eigenvalues of an operator by introducing the heat kernel,

$$K(\mathbf{x}, \mathbf{x}'; t) = \sum_n u_n(\mathbf{x}) u_n(\mathbf{x}') e^{-\lambda_n t}. \qquad (3.27)$$

The heat kernel also satisfies a heat equation,

$$\left(\Delta - \frac{\partial}{\partial t} \right) K(\mathbf{x}, \mathbf{x}', t) = 0 \qquad (3.28)$$

with the initial condition $K(\mathbf{x}, \mathbf{x}', 0) = \delta(\mathbf{x}, \mathbf{x}')$. There is a large literature on properties of the heat kernel, some of which is summarised in book form by Gilkey (1984).

The two separate spacetime points are somewhat superfluous and an integrated form of the heat kernel will serve,

$$K(f, t) = \int d\mu(x) f(\mathbf{x}) K(\mathbf{x}, \mathbf{x}; t). \qquad (3.29)$$

This can also be written as an infinite trace,

$$K(f, t) = \mathrm{tr} \left(f e^{-\Delta t} \right). \qquad (3.30)$$

The case $K(1, t)$ will be written $K(t)$.

The behaviour of the eigenvalues for large values of n is closely related to the behaviour of the heat kernel for small values of t. For very small t the heat kernel is concentrated in the immediate vicinity of $\mathbf{x} = \mathbf{x}'$ where the effects of curvature should be negligible. We would therefore expect that K is very close to the heat kernel in flat D-dimensional space. For a scalar field,

$$K(\mathbf{x}, \mathbf{x}', t) \sim \left(\frac{1}{2\pi t}\right)^{D/2} \exp\left(-(\mathbf{x} - \mathbf{x}')^2/4t\right). \tag{3.31}$$

Consequently, in four dimensions with volume Ω,

$$K(t) \sim \frac{\Omega}{4\pi^2} t^{-2}. \tag{3.32}$$

This, together with equation (3.27), is responsible for the statement that $\lambda_n \sim n^{1/2}$ in the previous section.

Quite generally, Gilkey has shown that the function $K(f, t)$ has an asymptotic expansion,

$$K(f, t) \sim t^{-D/2} \sum_{N=0}^{\infty} B_{N/2}(f) t^{N/2} \tag{3.33}$$

in D dimensions. The usefulness of this result follows from the fact that it can also be shown that the B coefficients depend only on the function f and the geometry and the operator,

$$B_N = (4\pi)^{-D/2} \int_M b_N(f) + (4\pi)^{-D/2} \int_{\partial M} c_N(f). \tag{3.34}$$

Explicit forms of $b_N(f)$ and $c_N(f)$ are known for small N. They are tabulated in appendix A.

The heat kernel contains similar information to a generalised ζ-function, defined by

$$\zeta(s) = \sum_n \lambda_n^{-s}. \tag{3.35}$$

If Δ only had a finite number of eigenvalues we could immediately use

$$\log \det \Delta = \sum_n \log \lambda_n = -\zeta'(0) \tag{3.36}$$

but in our case the sum diverges.

The ζ-function has an alternative definition by an integral transformation for $s > 2$,

$$\zeta(s) = \frac{1}{\Gamma(s)} \int_0^{\infty} K(t) t^{s-1} dt. \tag{3.37}$$

By separating off the integral for $t < 1$ and using the heat kernel expansion on this part, it is possible to express the ζ-function as

$$\zeta(s) = \frac{1}{\Gamma(s)} \left\{ \frac{B_{D/2}}{s} + \sum_{N \neq D} \frac{B_{N/2}}{(D-N)/2 + s} + w(s) \right\} \tag{3.38}$$

where $w(s)$ is a regular function of s. Analytic continuation can now be used to extend this result to $s < 2$. At $s = 0$, the Laurent expansion of $1/\Gamma(s) = s + \text{const} + \ldots$ cancels the pole term at the beginning and the function is regular with $\zeta(0) = B_{D/2}$.

The regular extension of the ζ-function provides a way to define the determinant of the operator. Consider the following definition:

$$\log \det \Delta = -\zeta'(0) - \zeta(0) \log \mu_R^2. \tag{3.39}$$

If we use analytic continuation from $s > 2$ to $s = 0$, then this definition is equivalent by equation (3.37) to

$$\log \det \Delta = -\frac{d}{ds} \left\{ \frac{1}{\Gamma(s)} \int_0^\infty dt\ t^{s-1} \mu_R^s \operatorname{tr} \left(e^{-\Delta t} \right) \right\}_{s=0}. \tag{3.40}$$

The shorter expression will usually be the one quoted and the regularisation method is accordingly known as ζ-function regularisation.

An alternative to ζ-function regularisation would be to replace the lower end of the integral in equation (3.40) by ϵ and take the limit $\epsilon \to 0$. It would be necessary to subtract terms $B_0\epsilon^{-2}$, etc., in order to obtain a finite result. This is called the 'proper-time cut-off' or 'Schwinger–DeWitt' method. Both methods are unambiguous apart from the arbitrariness of the coefficient of $\zeta(0)$, which must be fixed by experiment.

Example: The flat-space ζ-function.

Consider a massive scalar field with a wave operator $\Delta = -\nabla^2 + m^2$ (Planck's constant has been absorbed into the mass). The eigenvalues of this operator are $k^2 + m^2$, with the wavevector k. Therefore the ζ-function in a volume Ω is given by

$$\zeta(s) = \Omega \int \frac{d^4k}{(2\pi)^4} \left(k^2 + m^2 \right)^{-s}. \tag{3.41}$$

This integral diverges for $s < 2$. Introducing $x = |k|^2/m^2$,

$$\zeta(s) = \frac{\Omega}{16\pi^2} m^{4-2s} \int_0^\infty dx\ x(x+1)^{-s} \tag{3.42}$$

$$= \frac{\Omega}{16\pi^2} \frac{m^{4-2s}}{(s-2)(s-1)}. \tag{3.43}$$

This expression can be analytically continued to a function with poles at $s = 1$ and $s = 2$. In particular,

$$\zeta(0) = \frac{\Omega}{32\pi^2} m^4 \tag{3.44}$$

$$\zeta'(0) = -\frac{\Omega}{32\pi^2} \left(m^4 \log m^2 - \tfrac{5}{2} m^4 \right). \tag{3.45}$$

The effective action introduces an important feature of quantum field theories, namely anomalies, where quantum effects break a symmetry of the classical theory. When these were first discovered it was suspected that they were simply a feature of bad regularisation techniques that would eventually go away. However, a great amount of attention has been paid to anomalies, particularly because of their role in string theories, so that now their presence is seen as necessary and is relatively well understood.

An example of a classical symmetry subject to anomalies is conformal invariance under local rescaling of the metric, $g_{ab} \rightarrow \Omega^2 g_{ab}$. For a constant rescaling the eigenvalues $\lambda_n \rightarrow \Omega^{-2}\lambda_n$, but the definition of the path integral measure implies that this is equivalent to a change of renormalisation scale $\mu_R \rightarrow \Omega\mu_R$.

An analysis in appendix A shows that, as a consequence of the conformal anomaly, $\zeta(0)$ determines the trace of the stress–energy tensor,

$$\langle T^a{}_a \rangle = -\hbar \frac{b_2}{16\pi^2}. \tag{3.46}$$

The scalar function b_2 depends on the geometry.

This trace can be used to find the entire stress–energy tensor for a conformally invariant scalar field in spaces with large symmetry groups. If the only invariant tensor is g_{ab}, then $\langle T_{ab} \rangle = \frac{1}{4}\langle T^a{}_a \rangle g_{ab}$. In de Sitter space with cosmological constant Λ, for example,

$$\langle T_{ab} \rangle = \hbar g_{ab} \Lambda^2 / 2840\pi^2. \tag{3.47}$$

The natural interpretation of this term is a quantum correction to the cosmological constant.

3.5   Generating functions

A powerful technique in quantum field theory relies on introducing an external source $J(\mathbf{x})$, enabling operator expectation values to be derived from the vacuum amplitude. This is particularly important for interacting field theories, when they come to be analysed perturbatively.

A generating function $W[J]$ is defined by

$$e^{-W[J]/\hbar} = \int d\mu[\phi]\, e^{-I_J[\phi]/\hbar} \tag{3.48}$$

where

$$I_J[\phi] = I[\phi] + \int_{\mathcal{M}} d\mu\, J\phi. \tag{3.49}$$

It helps to change notation at this point and write ϕ^i instead of $\phi(x)$. Points are now labelled by the index i and a repeated index or a scalar product is used to indicate a volume integral,

$$J_i\phi^i = J \cdot \phi = \int_{\mathcal{M}} d\mu \, J(x)\phi(x). \tag{3.50}$$

Propagators can be obtained from the generating functional by differentiation,

$$\frac{\delta}{\delta J_i} J_j \phi^j = \phi^i. \tag{3.51}$$

Differentiating through the path integral gives

$$\langle \phi^i \rangle = \frac{\delta}{\delta J_i} W[J]. \tag{3.52}$$

Differentiating again,

$$\langle \phi^i \phi^j \rangle = \frac{\delta}{\delta J_i} \frac{\delta}{\delta J_j} W[J]. \tag{3.53}$$

When $J = 0$ this expectation value defines a propagator for the full theory. Consider the free field for example,

$$I_J[\phi] = \tfrac{1}{2}\phi^i \Delta_{ij} \phi^j + J_i \phi^i. \tag{3.54}$$

The generating function can be evaluated by completing the square in the action. The inverse of the operator Δ defines the Riemannian propagator G_E. The action can then be written

$$I_J[\phi] = \tfrac{1}{2}\{(\phi + J \cdot G_E) \cdot \Delta \cdot (\phi + G_E \cdot J)\} - \tfrac{1}{2}\{J \cdot G_E \cdot J\}. \tag{3.55}$$

This shift of variable allows us to evaluate the integral as before,

$$W[J] = \tfrac{1}{2}\hbar \log \det \Delta + \tfrac{1}{2} J \cdot G_E \cdot J. \tag{3.56}$$

We can now identify $\langle \phi^i \phi^j \rangle$ at $J = 0$ with G_E^{ij} by equation (3.53).

3.6   The effective action

An external source J induces non-vanishing background fields $\phi_c = \langle \phi(\mathbf{x}) \rangle$. It ought to simplify matters to work directly with ϕ_c and eliminate J. The result is the effective action $\Gamma[\phi]$ and a variational principle for the field expectation value.

If there are boundaries present, then $\Gamma[\phi]$ also depends on the boundary values $\widehat{\phi}$. When $J(\mathbf{x})$ vanishes $\Gamma[\phi]$ is the logarithm of the transition amplitude.

An important property of the effective action is that it generates equations for ϕ_c in the same way that the action generates the classical field equations. Thus

$$\frac{\delta\Gamma[\phi]}{\delta\phi^i} = \frac{\delta W[\phi]}{\delta\phi^i} - \frac{\delta J_j}{\delta\phi^i}\phi^j - J_j\frac{\delta\phi^j}{\delta\phi^i} = -J_i \qquad (3.58)$$

at $\phi = \phi_c$. When the source J vanishes, then the variation of the effective action vanishes at $\phi = \phi_c$.

A useful way of approximating the path integral is to use a saddle point method, where fields are expanded about backgrounds ϕ_s. It is convenient to make these solutions agree on the boundary with any boundary conditions. Replacing ϕ by $\phi_s + \phi_q$ in the action and expanding in powers of ϕ_q gives

$$I_J[\phi_s + \phi_q] = I_J[\phi_s] + I^{(1)}[\phi_s] + I^{(2)}[\phi_s] + \dots. \qquad (3.59)$$

The linear term depends on the classical field equations with the source included,

$$I^{(1)}[\phi_s] = F_J[\phi_s] \cdot \phi_q. \qquad (3.60)$$

The quadratic term has the form

$$I^{(2)}[\phi_s] = \tfrac{1}{2}\phi_q \cdot \Delta[\phi_s] \cdot \phi_q. \qquad (3.61)$$

These terms are sufficient to get terms in the perturbation series up to order $\hbar$.

The path integral to this order reads

$$Z[\widehat{\phi}, J] \sim e^{-I_J[\phi_s]} \int d\mu[\phi_q]\, e^{-(I^{(1)}+I^{(2)})/\hbar}. \qquad (3.62)$$

This can be evaluated as in the free case, resulting in

$$W[J] = I_J[\phi_s] + \tfrac{1}{2}F_J \cdot \Delta^{-1} \cdot F_J(\mathbf{x}') + \tfrac{1}{2}\hbar \log \det \Delta. \qquad (3.63)$$

We eliminate F_J by differentiating with respect to J, recalling the definition of ϕ_c and that F_J is linear in J,

$$\phi_c = \phi_s + \int d\mu(\mathbf{x}')\, \Delta^{-1}(\mathbf{x}, \mathbf{x}')F_J(\mathbf{x}') + O(\hbar). \qquad (3.64)$$

Choose $\phi_s = \phi_c$, then $F_J = O(\hbar)$, and we obtain

$$\Gamma[\phi] = I[\phi] + \tfrac{1}{2}\hbar \log \det \Delta[\phi] + O(\hbar^2). \qquad (3.65)$$

This is usually referred to as the one-loop approximation. As before, divergences have to be removed to make sense of the result.

In this background field expansion, the higher-order terms can be represented diagrammatically by Feynman diagrams. These can be recovered by considering the higher-order terms in the action, $\exp(-I^{(3)}/\hbar)$, etc. However, these higher-order terms will not be considered further in this book.

Example: A self-interacting scalar field

Consider a scalar field whose potential is given by a function $V(\phi)$ containing mass terms and self-interactions. The Lagrangian is

$$\mathcal{L} = -\tfrac{1}{2}\nabla\phi \cdot \nabla\phi - V(\phi). \tag{3.66}$$

The classical field equations are given by

$$F_J = -\nabla^2\phi + V'(\phi) + J \tag{3.67}$$

and the second-order variation about the solution ϕ gives a fluctuation operator,

$$\Delta = -\nabla^2 + m^2(\phi) \tag{3.68}$$

where $m^2(\phi) = V''(\phi)$. When ϕ is constant the quantum correction to the effective action will be similar to the result for a free field, but with the mass replaced by a function of ϕ. In flat space the result for a massive field was given in equation (8.50). If we define V^{eff} by $\Gamma = \Omega V^{\text{eff}}$, then

$$V^{\text{eff}}(\phi) = V(\phi) + \frac{\hbar}{64\pi^2}m(\phi)^4 \log m^2(\phi)/\mu_R^2. \tag{3.69}$$

The result of calculating the effective action is a function that depends on the mass scale μ_R. According to the correspondence principle, there should be a regime in which the classical theory is a good approximation, giving us access to the various parameters which appear there.

With quantum field theory, this simple picture is complicated by the fact that an extra process of regularisation has been applied between the classical and the final theory. This prevents an unambiguous determination of the classical parameters, and the best values to use depend on the experiments which are being considered. This is not to say that there is necessarily a larger number of parameters needed for the quantum theory. Theories with the same number of parameters in the quantum theory as in the classical theory are called renormalisable. However, when comparing different parametrisations it may be necessary to introduce an extra parameter, called the renormalisation scale.

The ζ-function regularisation scheme is unusual in that it removes divergences without directly subtracting any terms. In most schemes, divergent terms have to be discarded. For a renormalisable theory, this can be achieved by introducing divergent terms into the classical, often called 'bare', parameters. The bare parameters diverge to cancel the quantum divergences, leaving a finite theory.

The term in the effective action at one loop which depends on the renormalisation scale is given in terms of $\zeta(0)$,

$$\mu_R \frac{d\Gamma}{d\mu_R} = -\hbar \frac{1}{32\pi^2} \int_{\mathcal{M}} d\mu \, b_2 \tag{3.70}$$

where the coefficient b_2 is a local function of the background fields. The theory is renormalisable at one loop if this term can be absorbed into the classical action $I(\phi)$. For example, the coefficient b_2 for a scalar field is given by the results in appendix A,

$$b_2 = \tfrac{1}{2}V''(\phi)^2 - \tfrac{1}{6}R\,V''(\phi) + \dots . \tag{3.71}$$

A general potential which leads to a renormalisable theory at one loop would therefore be

$$V(\phi) = -\beta_0 \hbar^{1/2} R\phi - \chi_0 R\phi^2 + \tfrac{1}{2}m_0{}^2\phi^2 + \hbar^{-1/2}g_0\phi^3 + \hbar^{-2}\lambda_0\phi^4. \tag{3.72}$$

The μ dependence of the quantum correction can be absorbed into the potential $V(\phi)$ by defining new parameters χ, m and so on. Thus,

$$\lambda = \lambda_0 - \frac{9\hbar}{4\pi^2}\lambda_0{}^2 \log \mu_R^2 \tag{3.73}$$

If the measurable parameter is λ, then the dependence on μ_R can be absorbed. However, different experiments may measure different combinations of λ_0 and μ_R, and it is often convenient to consider how parameters vary at fixed values of λ_0.

3.7   Spontaneous symmetry breaking

Modern particle physics gives a very important role to models with underlying symmetries. Many properties of the elementary particles appear to follow from their symmetry properties, but with a subtle catch. The symmetry is often not a symmetry of the lowest-energy state of the system. In this section we will construct the effective action for a model of this type.

The simple example consists of a scalar field with the double-well potential figure 3.1. This potential has a symmetry transformation $\phi \mapsto -\phi$. There are two values of ϕ that minimise the energy and the symmetry is broken by particle excitations about either one.

In general the ground state of the field is a superposition of wavefunctions that are peaked about the minima. The ground-state expectation value

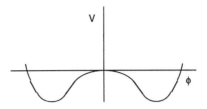

Figure 3.1 Scalar field potential.

The simple example consists of a scalar field with the double-well potential figure 3.1. This potential has a symmetry transformation $\phi \mapsto -\phi$. There are two values of ϕ that minimise the energy and the symmetry is broken by particle excitations about either one.

In general the ground state of the field is a superposition of wavefunctions that are peaked about the minima. The ground-state expectation value averages out to a value between the two minima but carries little information about the local behaviour of the field.

Previously the effective action was constructed by expressing an external source J in terms of the field expectation value ϕ_c. The leading-order behaviour of ϕ_c for the double-well potential is shown in figure 3.2 and clearly this is not an invertible function.

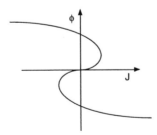

Figure 3.2 The field expectation value ϕ_c as a function of source J.

A better approach is to introduce an extra source $K(x, x')$ coupled to quadratic terms in the action. This source is able to force the quantum state to be concentrated near to the symmetric value of the field. The combination of $J(\mathbf{x})$ and $K(\mathbf{x}, \mathbf{x}')$ is sufficient to specify both the expectation value $\phi(\mathbf{x})$ and the propagator $G(\mathbf{x}, \mathbf{x}')$. The effective action can be defined by imposing

$J = 0$ and $K(\mathbf{x}, \mathbf{x}') = k(\mathbf{x})\delta(\mathbf{x}, \mathbf{x}')$. (If proper care is taken, these conditions are imposed as limits as the spatial volume goes to infinity.)

We begin with the generating function $Z[J, K]$ defined by a path integral,

$$Z[J, K] = \int d\mu[\phi] e^{-I_{JK}[\phi]/\hbar} \tag{3.74}$$

where, using the previous compact notation,

$$I_{JK}[\phi] = I[\phi] + J \cdot \phi + \tfrac{1}{2}\phi \cdot K \cdot \phi. \tag{3.75}$$

Expectation values for the field ϕ_c and the propagators are generated from $W = -\hbar \log Z$,

$$\phi_c^i = \frac{\delta W[J, K]}{\delta J_i} \tag{3.76}$$

$$\tag{3.77}$$

$$\hbar G^{ij} = 2\frac{\delta W[J, K]}{\delta K_{ij}} - \frac{\delta W[J, K]}{\delta J_i}\frac{\delta W[J, K]}{\delta J_j}. \tag{3.78}$$

These are inverted to give $J[\phi, G]$ and $K[\phi, G]$. The composite effective action is then defined by

$$\Gamma[\phi, G] = W[\phi, G] - J \cdot \phi - \phi \cdot K \cdot \phi - \hbar \operatorname{tr}(GK). \tag{3.79}$$

Variation of the composite action yields the equations

$$\frac{\delta \Gamma}{\delta \phi^i} = -J^i - (K \cdot \phi)^i \tag{3.80}$$

$$\frac{\delta \Gamma}{\delta G^{ij}} = -\tfrac{1}{2}K_{ij}. \tag{3.81}$$

If $K = 0$ then the second equation can be solved for G as a functional of ϕ. Substituting this back into the composite action would give the previous effective action. In the double-well potential we take $J = 0$ and $K = kI$ and solve for G. This also gives an effective action $\Gamma[\phi]$.

The saddle point method can be used to obtain an asymptotic series for the effective action. Set $\phi = \phi_s + \phi_q$. The expansion proceeds as before, but with F_J replaced by F_{JK} and Δ by $\Delta + K$. We obtain

$$\Gamma[\phi, G] = I[\phi] + \tfrac{1}{2}\hbar \log \det(\Delta + K) - \tfrac{1}{2}\hbar \operatorname{tr}(GK) + O(\hbar^2). \tag{3.82}$$

In addition, it follows from the definition of G that

$$G = (\Delta + K)^{-1} + O(\hbar). \tag{3.83}$$

This allows the elimination of K to give

$$\Gamma[\phi, G] = I[\phi] - \tfrac{1}{2}\hbar \log \det G - \tfrac{1}{2}\hbar \operatorname{tr}(1 - G\Delta) + O(\hbar^2). \qquad (3.84)$$

The final form of the effective action as a function of ϕ follows from equation (3.82) by imposing $J = 0$ and $K = kI$. During the derivation we had the equation $F_{JK} = O(\hbar)$. This equation fixes k (and also G), as a functional of ϕ.

Example: Explicit calculation for the scalar field.

We will calculate the effective action for a constant scalar field with classical potential

$$V(\phi) = -\tfrac{1}{2}\mu^2\phi^2 + \tfrac{1}{4}\lambda\phi^4. \qquad (3.85)$$

Introduction of the quadratic source changes the field equations to

$$F_{JK} = -\nabla^2\phi + V'(\phi) - J + \int d\mu(x')K(x, x')\phi(x'). \qquad (3.86)$$

With $J = 0$ and $K = kI$ the condition that $F_{JK} = O(\hbar)$ gives

$$k = \mu^2 - \lambda\phi^2 + O(\hbar). \qquad (3.87)$$

Therefore

$$G^{-1} = -\nabla^2 + V''(\phi) + k = -\nabla^2 + 2\lambda\phi^2. \qquad (3.88)$$

If the problem with inverting the relation for the source had been ignored, we would have obtained the corresponding operator

$$\Delta = -\nabla^2 + V''(\phi) = -\nabla^2 - \mu^2 + 3\lambda\phi^2. \qquad (3.89)$$

This latter operator is not even positive definite and there would be ambiguities in making sense out of the determinant.

4

Gauge theories

Local symmetry is a very important component of relativistic field theory. We have already seen some of the effects of Lorentz symmetry in imposing the classification of fields into scalars, vectors and spinors and restricting the possible Lagrangians. Now is the time to consider the consequences of internal gauge symmetry.

An internal symmetry is a transformation of the fields by the action of a group. Some important groups are given in table 4.1. In a gauge transformation, $\phi(\mathbf{x}) \mapsto \mathbf{U}(\mathbf{x})\phi(\mathbf{x})$, where $\mathbf{U}$ is a representation of the group and $\phi(\mathbf{x})$ is a field.

The quantisation of gauge theories should be set up in a way which respects the gauge symmetries if we wish to ensure that the physical consequences of the symmetries are preserved. This complicates any perturbative expansion of the path integral, which usually requires some form of gauge fixing. The effects of gauge fixing have to be removed by introducing new fields. Here, we will use a technique known as the BRST approach (named after Becchi–Rouet–Stora–Tyutin).

The need for extra fields was first realised by Feynman (1963), and developed in detail by DeWitt (1967b,c). Many of the important issues raised in quantising gauge theories are presented in an entertaining way by Coleman (1985). A far more detailed account of the BRST approach is given by Henneaux and Teitelboim (1992).

Table 4.1 Classical gauge groups.

GL(N,R)	$N \times N$ invertible real matrices	
SL(N,R)	$N \times N$ invertible real matrices	$\det \mathbf{M} = 1$
U(N)	$N \times N$ unitary matrices	$\mathbf{U}^\dagger \mathbf{U} = \mathbf{1}$
SU(N)	$N \times N$ unitary matrices	$\det \mathbf{U} = 1$
O(N)	$N \times N$ orthogonal matrices	$\mathbf{O}\mathbf{O}^T = \mathbf{1}$
SO(N)	$N \times N$ orthogonal matrices	$\det \mathbf{O} = 1$

4.1   Gauge fixing

By way of example we will consider how to quantise scalar or spinor electromagnetism, which has the simple gauge symmetry group U(1). One way in which the symmetry presents itself is that solutions of the field equations should transform into other solutions under the gauge transformations

$$\psi(\mathbf{x}) \mapsto e^{i\theta(x)}\psi(\mathbf{x}). \tag{4.1}$$

The transformation is local in the sense that the parameter θ depends upon the coordinate $\mathbf{x}$.

Ordinary covariant derivatives ∇ in the field equations have to be replaced by gauge covariant derivatives $\mathbf{D}$, which include spacetime vector gauge connections $g\mathbf{A}$,

$$\mathbf{D} = \nabla - ig\mathbf{A}. \tag{4.2}$$

Under the gauge transformation we require that gauge covariant derivatives of ψ should transform in the same way as ψ. The connection vector ensures this by transforming in a way that cancels derivatives of the gauge parameter,

$$g\mathbf{A} \mapsto g\mathbf{A} + \nabla\theta. \tag{4.3}$$

We will identify $\mathbf{A}$ as the relativistic field of the photon and relate the coupling constant g to the electronic charge unit e and the permeability of free space μ_0 by $g^2 = e^2\mu_0/\hbar^2$. This appearance of a vector boson field is typical of gauge theories and an extremely attractive feature.

The dynamical properties of the photon field have still to be specified. The field strength tensor $\mathbf{F}$,

$$F_{ab} = \nabla_a A_b - \nabla_b A_a \tag{4.4}$$

remains unchanged under the action of the U(1) gauge group. Therefore the Lagrangian,

$$\mathcal{L}_g = -\tfrac{1}{4}F_{ab}F^{ab} \tag{4.5}$$

is invariant under both spacetime symmetries and gauge transformations.

The relativistic wave equation for the vector potential can be obtained from the functional derivative,

$$\frac{\delta I}{\delta A_a} = \Delta_a{}^b A_b. \tag{4.6}$$

The wave operator appearing here is

$$\Delta_a{}^b = -\delta_a{}^b \nabla_c \nabla^c + \nabla_a \nabla^b + R_a{}^b \tag{4.7}$$

after the Ricci identity has been used. The vanishing of the functional derivative is equivalent to the vacuum Maxwell equation $\nabla \cdot \mathbf{F} = 0$.

The difficulty that arises in using a path integral to generate propagators can now be clearly seen. The problem is that the operator Δ has vanishing determinant and no inverse, because $\Delta_a{}^b \theta_{,b} = 0$. This is no accident, but a consequence of the gauge invariance.

The resolution of the problem is to remove gauge ambiguity in some way. A term called the gauge fixing term is added to the Lagrangian for this purpose. For example,

$$\mathcal{L}_{gf} = -\frac{\alpha}{2}(\boldsymbol{\nabla} \cdot \mathbf{A})^2. \qquad (4.8)$$

The operator for the gauge field becomes now

$$\Delta_a{}^b = -\delta_a{}^b \nabla_c \nabla^c + (1 - \alpha)\nabla_a \nabla^b + R_a{}^b. \qquad (4.9)$$

The Riemannian propagator is the inverse of this operator. For $\alpha = 1$, the propagator is said to be in the 'Feynman gauge'. The limit $\alpha \to \infty$ results in a purely tranverse propagator and this is the 'Landau gauge'.

4.2   BRST symmetry

Adding gauge fixing terms to the action is a rather arbitrary proceedure and something has to be done to ensure that the theory is not fundamentally changed in the process. This means ensuring that nothing physical depends on the choice of gauge fixing term. In practice this can be done by introducing a BRST symmetry.

In the BRST approach extra fields are added to the theory. The extra fields are called ghosts and they are always taken to vanish when any physical process is considered. The ghost fields cancel extra degrees of freedom introduced by the gauge fixing and this means that some have to be anticommuting fields. The BRST symmetry is obtained by replacing the gauge parameter with a new field and adding extra terms to the action,

$$\mathbf{s}\,\psi = igc\psi \qquad \mathbf{s}\,\mathbf{A} = \boldsymbol{\nabla}c. \qquad (4.10)$$

The gauge-invariant terms in the action inherit BRST symmetry, but the variation of the gauge fixing term is non-vanishing,

$$\mathbf{s}\mathcal{L}_{gf} = -\alpha(\boldsymbol{\nabla} \cdot \mathbf{A})\nabla^2 c. \qquad (4.11)$$

The simplest way in which to cancel this variation is to add a term

$$\mathcal{L}_{gh} = \bar{c}\nabla^2 c \qquad (4.12)$$

and then define

$$\mathbf{s}\,c = 0 \qquad \mathbf{s}\,\bar{c} = \alpha(\boldsymbol{\nabla} \cdot \mathbf{A}). \qquad (4.13)$$

The total action is then

$$I = - \int d\mu(x) \left(\mathcal{L}_g + \mathcal{L}_{gf} + \mathcal{L}_{gh} \right). \tag{4.14}$$

There is no coupling between the ghosts and the photons, and the ghosts have no effect on the theory. However, this is only the case for very simple theories, like electromagnetism.

An important restriction for BRST transformations is to have them be nilpotent, in other words $\mathbf{s}^2 = 0$. This is important for transforming the measure in the path integral. Under a BRST transformation on the fields the Feynman measure changes by a Jacobian factor,

$$\det(\mathbf{1} + \mathbf{s}). \tag{4.15}$$

We have seen already that such determinants can be defined for operators, and that they satisfy $\det \Delta = \exp \operatorname{tr} \log \Delta$. If $\mathbf{s}$ is nilpotent, then

$$\det(\mathbf{1} \pm \mathbf{s}) \simeq \mathbf{1} \pm \operatorname{tr} \mathbf{s}. \tag{4.16}$$

A sign $\simeq$ is used to denote equality up to terms appearing in the regularisation procedure. Noting that $(\mathbf{1}+\mathbf{s})(\mathbf{1}-\mathbf{s}) \simeq \mathbf{1}$ we see that $\operatorname{tr} \mathbf{s} \simeq 0$ and the Jacobian factor is equal to unity. The simplest path integral measure is therefore invariant under BRST transformations.

The existing BRST transformations are nilpotent only when the ghost fields satisfy $\nabla^2 c = 0$. This restriction can be removed by introducing a new commuting field b, called an antighost, and then the complete set of transformations is

$$\mathbf{s}\,\mathbf{A} = \boldsymbol{\nabla} c \quad \mathbf{s}\,c = 0 \quad \mathbf{s}\,\bar{c} = ib \quad \mathbf{s}\,b = 0. \tag{4.17}$$

The gauge fixing Lagrangian which is invariant under this symmetry is

$$\mathcal{L}_{gf} = -ib(\boldsymbol{\nabla} \cdot \mathbf{A}) - \frac{1}{2\alpha}b^2 \tag{4.18}$$

from which it is clear that in the Landau gauge b resembles a Lagrange multiplier. When $b = -i\alpha \boldsymbol{\nabla} \cdot \mathbf{A}$, then the gauge fixing term and symmetry revert to the previous case.

The gauge fixing and ghost Lagrangians can also be written in the form

$$\mathcal{L}_{gf} + \mathcal{L}_{gh} = s \left(-\bar{c}\boldsymbol{\nabla} \cdot \mathbf{A} + \frac{i}{2\alpha}\bar{c}b \right). \tag{4.19}$$

In this form the BRST invariance of the action follows trivially from the nilpotency.

BRST symmetry can be applied equally well to gauge symmetry in phase space. By way of example we take the action for a relativistic particle in a form independent of the parametrisation of the trajectory,

$$S = \int_0^1 (\dot{x}^a p_a - NH)\, d\tau \qquad (4.20)$$

where $H = \frac{1}{2}g^{ab}p_a p_b + \frac{1}{2}m^2$. The variable $N(\tau)$ measures the rate of change of proper time with respect to the parameter τ along the world-line. The field equations,

$$\dot{x}^a = N\frac{\partial H}{\partial p_a} \quad \dot{p}_a = -N\frac{\partial H}{\partial x^a} \quad H = 0 \qquad (4.21)$$

imply geodesic motion.

The action has a gauge invariance under reparametrisations of the path $\tau \equiv \tau(\tau')$. The infinitesimal transformations are $\delta x = c\dot{x}$, $\delta p = c\dot{p}$ and $\delta N = \dot{c}$. We would like to replace this gauge symmetry by a BRST symmetry, particularly one which replaces time derivatives by momenta,

$$sx^a = c\frac{\partial H}{\partial p_a} \quad sp_a = -c\frac{\partial H}{\partial x^a} \quad sN = \rho. \qquad (4.22)$$

We will choose a gauge fixing term

$$L_{gf} = b\dot{N} \qquad (4.23)$$

where b is an antighost field. The ghost action must make up for the fact that the BRST variation of N is ρ instead of $\dot{c}$,

$$L_{gh} = \overline{\rho}(\dot{c} - \rho) + \overline{c}\dot{\rho}. \qquad (4.24)$$

The total action $L = \dot{x}^a p_a - NH + L_{gf} + L_{gh}$ is BRST invariant with

$$sc = s\rho = sb = 0 \quad s\overline{\rho} = -H \quad s\overline{c} = -b. \qquad (4.25)$$

Example: The free-particle propagator in flat spacetime

The covariant path integral, with $Q = \{x, N, \rho, c\}$ and $P = \{p, b, \overline{\rho}, \overline{c}\}$, takes the form

$$G(x_0, x_1) = \int d\mu[Q, P]\, \mathcal{P}(x_1|x_0)e^{iS[Q,P]/\hbar}. \qquad (4.26)$$

The ghosts decouple and integration over b leaves only a dependence on the elapsed time,

$$T = \int_0^1 N(\tau)d\tau. \qquad (4.27)$$

The amplitude becomes

$$G(x_0, x_1) = \int dT \, d\mu[x, p] \exp\left(\frac{i}{\hbar} \int_0^T (p^a \dot{x}_a - H) dt\right). \qquad (4.28)$$

The exponent can be rewritten,

$$-i \int_0^T (x^a \dot{p}_a + H) dt + (\mathbf{p}_1 \cdot \mathbf{x}_1 - \mathbf{p}_0 \cdot \mathbf{x}_0). \qquad (4.29)$$

We can perform the path integral over $x(t)$ for $0 < t < T$ which makes the momentum p_a and the Hamiltonian H independent of t,

$$G(x_0, x_1) = \int dT \, d\mu(p) \exp\{i[p_a(x_1 - x_0)^a - T(p^2 + m^2)]/\hbar\}. \qquad (4.30)$$

This is the heat kernel form of the propagator in flat space originally derived by Schwinger. The range of the T integral can be chosen to run from 0 to ∞. This results in the Green function,

$$G(x_0, x_1) = \int_C d\mu(p) \frac{\exp\{i\mathbf{p} \cdot (\mathbf{x}_1 - \mathbf{x}_0)/\hbar\}}{(p^2 + m^2)} \qquad (4.31)$$

whose form depends on the contour chosen for the E integration.

4.3 Consequences of BRST symmetry

Path integrals can be used to quantise gauge theory, defining propagators and effective actions as was done for scalar fields. Interacting theories can be analysed using perturbation theory, as before. There is one additional tool available to the study of gauge theories, namely the BRST symmetry.

The generating path integral for the pure gauge theory now includes ghosts,

$$Z[\mathbf{J}, \bar{\theta}, \theta] = \int d\mu[\mathbf{A}, c, \bar{c}] \exp(-I_{J, \bar{\theta}, \theta}[\mathbf{A}, c, \bar{c}]) \qquad (4.32)$$

if the antighost is neglected, where

$$I_{J, \bar{\theta}, \theta} = -\int d\mu(x) \left(\mathcal{L}_g + \mathcal{L}_{gf} + \mathcal{L}_{gh} + J^a A_a + \bar{\theta} c + \bar{c} \theta\right). \qquad (4.33)$$

Expectation values of operators are defined by the path integral,

$$\langle \mathbf{F} \rangle = \frac{1}{Z[\mathbf{J}, \bar{\theta}, \theta]} \int d\mu[\mathbf{A}, c, \bar{c}] \, F \, \exp(-I_{J, \bar{\theta}\theta}). \qquad (4.34)$$

Green functions are constructed from $W = -\log Z$. Electron or other matter fields should also be included in the action to build up a complete theory including interactions.

A fundamental requirement of this approach is that the path integral measure must be invariant under the BRST symmetry. This is equivalent to the statement

$$\int d\mu[\mathbf{A}, c, \bar{c}] \, sF[\mathbf{A}, c, \bar{c}] = 0 \tag{4.35}$$

for any functional $F[\mathbf{A}, c, \bar{c}]$, or in particular

$$\int d\mu[\mathbf{A}, c, \bar{c}] \, (s \, I_{J,\bar{\theta},\theta}) \exp(-I_{J,\bar{\theta},\theta}) = 0. \tag{4.36}$$

Because of BRST invariance we have

$$s \, I_{J,\bar{\theta},\theta} = - \int d\mu(x) \left(J^a s A_a + \bar{\theta} s c + s\bar{c}\theta \right). \tag{4.37}$$

If we recall the definitions of the classical field expectation values, then

$$\int d\mu(x) \left(J^a \langle s A_a \rangle + \bar{\theta}\langle sc \rangle + \langle s\bar{c}\rangle\theta \right) = 0. \tag{4.38}$$

But the sources are related to an effective action by

$$J^a = \frac{\delta\Gamma}{\delta A_a} \qquad \theta = \frac{\delta\Gamma}{\delta\bar{c}} \qquad \bar{\theta} = -\frac{\delta\Gamma}{\delta c}. \tag{4.39}$$

Consequently $s\Gamma[\mathbf{A}, c, \bar{c}] = 0$, if we define $s\langle \mathbf{A} \rangle = \langle s\mathbf{A} \rangle$. This invariance of the effective action contains all of the physical consequences of the original gauge invariance.

When matter fields are included the BRST action becomes non-linear and the condition $s\langle \psi \rangle = \langle s\psi \rangle$, for example, is no longer obvious. In this case it is possible to introduce extra sources $\mathbf{u}$, v, w into the action coupled to $s\mathbf{A}$, sc, $s\psi$, thus

$$\frac{\delta\Gamma}{\delta u}\frac{\delta\Gamma}{\delta \mathbf{A}} + \frac{\delta\Gamma}{\delta v}\frac{\delta\Gamma}{\delta \bar{c}} + \frac{\delta\Gamma}{\delta \psi}\frac{\delta\Gamma}{\delta w} + \frac{\delta\Gamma}{\delta \bar{w}}\frac{\delta\Gamma}{\delta \bar{\psi}} = 0. \tag{4.40}$$

Equations of this type are called Ward–Slavnov–Taylor identities.

Example: The one-loop photon effective action as a function of the background metric

It is convenient to use differential p-forms Λ_p, with exterior derivative d, hodge star operator $*$ and $\delta = *d*$. The vector potential has the differential properties of a 1–form field. Viewed this way, the Maxwell operator can be written as δd.

The gauge–fixed operator

$$\Delta = \delta d + \alpha d\delta = \Delta_h + (\alpha - 1)d\delta, \tag{4.41}$$

where Δ_h is the Hodge–DeRahm Laplacian. According to the Hodge decomposition theorem,

$$\Lambda_1 = d\Lambda_0 \oplus \delta\Lambda_2 \oplus H_1 \qquad (4.42)$$

where H_1 is the space of harmonic forms, which we shall take to be empty for simplicity. The restriction of Δ to a space Λ will be denoted by $\Delta(\Lambda)$.

The orthogonality of the decomposition allows us to infer that

$$\det \Delta(\Lambda_1) = \det \Delta(d\Lambda_0) \det \Delta(\delta\Lambda_2) \qquad (4.43)$$

The first term can be interpreted as the contribution from pure gauge modes and the second from the transverse degrees of freedom.

Suppose that $\eta \in \Lambda_0$, and that η is an eigenfunction of Δ_h with eigenvalue λ. Then,

$$\Delta d\eta = d\Delta_h\eta + (\alpha - 1)d\delta d\eta = \alpha\lambda d\eta. \qquad (4.44)$$

Consequently, we can write

$$\det \Delta(d\Lambda_0) = \det(\alpha\Delta(\Lambda_0)). \qquad (4.45)$$

In the gauge-fixed theory we also have anticommuting ghosts $c, \bar{c}$, with ghost operator $\Delta(\Lambda_0)$ and the antighost b, with 'operator' $\alpha^{-1}\mathbf{1}$. The result of evaluating the path integral over photon and ghost fields, with $Z[g] = \exp(-\Gamma[g])$, is therefore

$$\Gamma[g] = \tfrac{1}{2}\log \det \Delta(\Lambda_1) - \log \det \Delta(\Lambda_0) - \tfrac{1}{2}\log \det \alpha(\Lambda_0). \qquad (4.46)$$

From the decomposition,

$$\Gamma[g] = \tfrac{1}{2}\log \det \Delta(\delta\Lambda_2) - \tfrac{1}{2}\log \det \Delta(\Lambda_0) \qquad (4.47)$$

and all dependence upon the parameter α has disappeared. We are free to take $\alpha = 1$ in particular,

$$\Gamma[g] = \tfrac{1}{2}\log \det \Delta_h(\Lambda_1) - \log \det \Delta(\Lambda_0). \qquad (4.48)$$

This is often the most useful form of the effective action, though the penultimate expression shows most clearly that calculating the determinant of *only* the transverse degrees of freedom would give totally the wrong result.

4.4   Non-Abelian gauge theories

The standard models of the fundamental forces of Nature, with the exception of gravity, are constructed around non-commutative symmetry groups. These groups are always Lie groups, where elements $\mathbf{U}$ near to the identity can be obtained by an exponential $\mathbf{U} = \exp(i\boldsymbol{\Lambda})$ with $\boldsymbol{\Lambda}$ belonging to the Lie algebra. A basis $\{\mathbf{T}^i\}$ for the Lie algebra satisfies

$$[\mathbf{T}^i, \mathbf{T}^j] = if^{ij}{}_k \mathbf{T}^k \tag{4.49}$$

with structure constants $f^{ij}{}_k$ that depend on the choice of basis. There is a natural metric on the Lie algebra, $g^{ij} = \text{tr}(\mathbf{T}^i \mathbf{T}^j)$, which can also be used to raise or lower indices.

The covariant derivative $\mathbf{D}$ is defined so that $\mathbf{D}\Phi^g = \mathbf{U}\mathbf{D}\Phi$ for a field which transforms under the gauge group to $\Phi^g = \mathbf{U}\Phi$. A field strength tensor can be defined from this derivative by

$$\mathbf{F}_{ab} = \frac{i}{g}[\mathbf{D}_a, \mathbf{D}_b]. \tag{4.50}$$

If the covariant derivative is decomposed into a spacetime and gauge connection part, $\mathbf{D} = \boldsymbol{\nabla} - ig\mathbf{A}$ where $\mathbf{A} = \mathbf{A}_i \mathbf{T}^i$, then

$$\mathbf{F}_{ab} = \nabla_a \mathbf{A}_b - \nabla_b \mathbf{A}_a - ig[\mathbf{A}_a, \mathbf{A}_b]. \tag{4.51}$$

Under a gauge transformation,

$$\mathbf{A}_a^g = \mathbf{U}\mathbf{A}_a\mathbf{U}^{-1} - ig^{-1}(\boldsymbol{\nabla}_a\mathbf{U})\mathbf{U}^{-1} \qquad \mathbf{F}_{ab}^g = \mathbf{U}\mathbf{F}_{ab}\mathbf{U}^{-1}. \tag{4.52}$$

The infinitesimal form of this transformation for the potential is

$$\delta_g \mathbf{A}_a = \mathbf{A}_a^g - \mathbf{A}_a = g^{-1}\mathbf{D}_a\boldsymbol{\Lambda}. \tag{4.53}$$

The field strength takes the adjoint representation of the group. For a field Φ in the adjoint representation,

$$\mathbf{D}_a\Phi = \nabla_a\Phi - ig[\mathbf{A}_a, \Phi]. \tag{4.54}$$

The gauge symmetry invites a natural choice of Lagrangian density for the gauge fields,

$$\mathcal{L}_A = -\tfrac{1}{4}\text{tr}(\mathbf{F}_{ab}\mathbf{F}^{ab}) = -\tfrac{1}{4}g^{ij}(F_{ab})_i(F^{ab})_j \tag{4.55}$$

and for the matter fields,

$$\mathcal{L}_\phi = -\tfrac{1}{2}(\mathbf{D}_a\Phi)^T(\mathbf{D}^a\Phi) - V(\Phi). \tag{4.56}$$

In the quantum theory we add a gauge fixing term and ghosts to set up perturbative expansions. This can be of the form

$$\mathcal{L}_{gf} = -\frac{\alpha}{2}\mathcal{F}_i\mathcal{F}^i \tag{4.57}$$

with $\mathcal{F}(\mathbf{A}, \Phi)$. The Lagrangian density of the ghost fields required by the BRST symmetry s would be

$$\mathcal{L}_c = \bar{c}^i s \mathcal{F}_i \tag{4.58}$$

corresponding to the transformations

$$s\Phi = igc_i \mathbf{T}^i \Phi \qquad s\,\mathbf{A} = \mathbf{D}c \tag{4.59}$$

$$sc_i = -\tfrac{1}{2} c_j c_k f^{jk}_{\ \ i} \qquad s\bar{c} = \alpha \mathcal{F} \tag{4.60}$$

Gauge symmetries are symmetries of the Lagrangian but need not be symmetries of a given quantum state, even if this state is the vacuum state. In the Higgs mechanism, the symmetry of the vacuum is broken by a scalar field which has a vacuum expectation value different from zero. This spontaneous breaking of a gauge symmetry is the basis of the unified models of particle interactions.

For example, an $O(N)$ invariant scalar model can be constructed from a potential

$$V(\Phi) = -\tfrac{1}{2}\mu^2 \Phi^T \Phi + \tfrac{1}{4}\lambda(\Phi^T \Phi)^2. \tag{4.61}$$

In the vacuum state $\langle \Phi \rangle = \phi$ where ϕ is a minimum of the potential $V(\phi)$. The set of these minima is called the vacuum manifold, in this case the sphere $|\phi| = \mu/\lambda^{1/2}$. The $O(N)$ symmetry is reduced to the $O(N-1)$ group which fixes ϕ. Scalar field fluctuations in the direction of the vacuum manifold are at constant potential and they are therefore massless. These are known as Goldstone modes. However, as the following shows, the Goldstone modes mix with the gauge fields and form massive vector bosons.

Example: The effective action for an O(N) Yang–Mills theory with a scalar Higgs field to one loop order

The effective action for a non-Abelian gauge theory can be found by expanding the fields about background fields $\mathbf{A}$ and ϕ with perturbations $\tilde{\mathbf{A}}$, $\tilde{\phi}$. A gauge fixing term also has to be added at this time. By a judicious choice of gauge fixing term, we can arrange to cancel some cross-terms between the fluctuations, which would otherwise have to be removed by diagonalising the operators that act on them. We use

$$\mathcal{F} = \mathbf{D}_a \tilde{\mathbf{A}}^a + i\alpha^{-1} g \phi^T T_i \tilde{\phi}\, T^i. \tag{4.62}$$

This choice is called the t'Hooft or R_ξ gauge fixing term.

The gauge fixing does not affect the invariance of the action under gauge transformations of the background fields. In addition, we also have the BRST symmetry transformations on the perturbed fields (4.60). What we cannot guarantee at this stage is a result that does not depend on the choice of gauge fixing term. Any dependence on the gauge fixing term would modify the effective field equations. However, it can be shown that if the effective action is truncated to order $\hbar$, then solutions to the effective field equations differ only at order $\hbar^2$. We shall be using Landau gauge, $\alpha \to \infty$. At higher orders in $\hbar$ the effective action has to be modified to remove any dependence on the gauge fixing terms (DeWitt 1964).

The ghost field c Lagrangian density reads

$$\mathcal{L}_c = \bar{c}^i s \mathcal{F}_i = \bar{c}^i (\mathbf{D} \cdot (\mathbf{D} - ig\tilde{\mathbf{A}})\delta_i^j - \alpha^{-1} g^2 \phi^T \mathbf{T}_i \mathbf{T}^j \Phi) c_j. \qquad (4.63)$$

The total Lagrangian $\mathcal{L}_{\text{tot}} = \mathcal{L}_A + \mathcal{L}_\phi + \mathcal{L}_{gf} + \mathcal{L}_{gh}$. If we expand the Lagrangian in the perturbation fields, after some effort we arrive at the quadratic part

$$\mathcal{L}_{\text{tot}} = -\tfrac{1}{2}\tilde{\mathbf{A}}\Delta_A\tilde{\mathbf{A}} - \tfrac{1}{2}\tilde{\phi}\Delta_\phi\tilde{\phi} - \tfrac{1}{2}c^T\Delta_c c - 2ig(\mathbf{D}\phi)^T \cdot \tilde{\mathbf{A}}\tilde{\phi} + 2ig\tilde{\mathbf{A}} \cdot (\mathbf{D}\phi)\tilde{\phi} + \mathcal{L}_{\text{int}}$$
$$(4.64)$$

where Δ_A, Δ_ϕ and Δ_c are the fluctuation operators, and $\mathcal{L}_{\text{int}}$ contains the higher-order interaction terms. We write

$$\Delta_A = -\delta_a{}^b\nabla^2 + (1-\alpha)\nabla_a\nabla^b + \mathbf{X}_A \qquad (4.65)$$

$$\Delta_\phi = -\nabla^2 + \mathbf{X}_\phi \qquad \Delta_c = -\nabla^2 + \mathbf{X}_c. \qquad (4.66)$$

For the vector and Higgs fields in the Landau gauge $\alpha \to \infty$,

$$\mathbf{X}_A = g^2 \phi^T \mathbf{T}_i \mathbf{T}^j \phi \delta_a{}^b + ig\,\text{tr}(\mathbf{T}_i \mathbf{F}_a{}^b \mathbf{T}^j) \qquad (4.67)$$

$$\mathbf{X}_\phi = 2\lambda\phi\phi^T \qquad \mathbf{X}_c = 0. \qquad (4.68)$$

Some of the fluctuating fields have developed masses, including some of the gauge bosons. These cause the gauge interactions to become short-range. This is the phenomenon known as spontaneous symmetry breaking.

We can identify these masses more accurately by introducing projection matrices $\mathbf{P}_\phi$ and $\mathbf{P}_A$, where $\mathbf{P}_\phi = |\phi|^{-2}\phi\phi^T$ is a projection along ϕ and $P_{A_i}{}^j = 2\,\text{tr}(\mathbf{P}_\phi \mathbf{T}_i \mathbf{T}^j \mathbf{P}_\phi)$. The complementary matrix $1 - \mathbf{P}_A$ projects onto the subgroup of symmetries that keep the background Higgs field ϕ fixed, $O(N-1)$.

For constant background fields the mass matrices become $\mathbf{X} = m^2\mathbf{P}$, where

$$m_\phi^2 = 2\lambda\phi^T\phi \qquad m_A^2 = \tfrac{1}{2}g^2\phi^T\phi. \qquad (4.69)$$

There are also massless gauge fields associated with the unbroken subgroup. The massless scalar fields and the ghost fields have been replaced by a single massive vector boson.

The one–loop effective action is given by

$$\Gamma^{(1)} = \tfrac{1}{2}\hbar \log \det \Delta_A + \tfrac{1}{2}\hbar \log \det \Delta_\phi - \hbar \log \det \Delta_{gh}. \qquad (4.70)$$

We can use the results for one-loop effective potential from the earlier section,

$$V^{(1)} = -\frac{\hbar}{64\pi^2} \left(m_\phi^4 \log \frac{m_\phi^2}{\mu_R^2} + 3(N-1)m_A^4 \log \frac{m_A^2}{\mu_R^2} \right) \qquad (4.71)$$

where μ_R is the renormalisation scale. The massless case with $\mu = 0$ and $\lambda = O(g^4)$ is known as the Coleman–Weinberg model,

$$V^{(1)} = \frac{3(N-1)\hbar}{256\pi^2} g^4 (\phi^T \phi)^2 \log \frac{\phi^T \phi}{\mu_R^2}. \qquad (4.72)$$

In this case the symmetry breaking is induced by the quantum corrections.

4.5 Gravity and gauge theory

Einstein's theory of gravity is based upon the premise that the fundamental equations should be covariant under local Lorentz transformations of the fields. The theory is therefore a prime example of a gauge theory. In this section we will see how far we can get using the gauge theory approach to obtain a quantum theory of gravity.

The action principle for gravity is as old as the theory of general relativity. The analogue of the field strength tensor is the Riemannian curvature tensor, but there is an important difference between the theory of gravity and other non-Abelian gauge theories: the dynamical variable in gravity is the metric rather than the connection. The action for gravity should be second order in derivatives of the metric; therefore the natural choice for the Lagrangian $\mathcal{L}$ is not the usual gauge theory action, but the Ricci scalar R instead,

$$\mathcal{L}_g = \frac{1}{2\kappa^2} R \qquad (4.73)$$

with κ a constant. If the spacetime manifold has a boundary then the action also has boundary terms.

An effective action can be constructed in just the same way as for any gauge theory. At one-loop order the dependence on renormalisation scale is determined as ever by heat kernel coefficients. Here an obvious problem with renormalisability arises due to quadratic curvature terms in the coefficients. This problem gets worse at each order in $\hbar$, and in general the effective action will not be renormalisable, but a finite set of new constants will enter at each order in the loop expansion. In this respect, Einstein's theory of gravity leads to a less satisfactory quantum theory than other gauge theories.

Perturbation theory is constructed by expanding the metric about a classical background metric g,

$$\widehat{g}_{ab} = g_{ab} + \gamma_{ab}. \qquad (4.74)$$

Consistency requires adding gauge fixing and ghost terms to the action. Gauge symmetry corresponds to coordinate transformations $\mathbf{x}' = \mathbf{x} + \mathbf{c}$. There will therefore be four gauge fixing conditions $\mathcal{F}_a(\mathbf{g}, \boldsymbol{\gamma})$, and a typical gauge fixing term will be of the form

$$\mathcal{L}_{gf} = -\alpha \widehat{g}^{ab} \mathcal{F}_a \mathcal{F}_b \tag{4.74}$$

The gauge symmetry gets replaced by BRST transformations,

$$s\, \gamma_{ab} = 2c_{(a;b)}. \tag{4.75}$$

The Lagrangian for the ghost field $\mathbf{c}$ becomes,

$$\mathcal{L}_{gh} = 2\overline{c}^a\, s\, \mathcal{F}_a \tag{4.76}$$

ensuring BRST invariance of the action.

The generating functional for gravity is now

$$Z[\mathbf{T}, \theta, \overline{\theta}] = \int d\mu[\widehat{\mathbf{g}}, \mathbf{c}, \overline{\mathbf{c}}] \, \exp(-I_{T,c,\overline{c}}[\widehat{\mathbf{g}}, \mathbf{c}, \overline{\mathbf{c}}]/\hbar), \tag{4.77}$$

with $\mathcal{L} = \mathcal{L}_g + \mathcal{L}_{gf} + \mathcal{L}_{gh}$ and

$$I_{T,c,\overline{c}}[\widehat{\mathbf{g}}, \mathbf{c}, \overline{\mathbf{c}}] = -\int_{\mathcal{M}} d\mu \, (\mathcal{L} + T^{ab}\widehat{g}_{ab} + \overline{\theta}^a c_a + \overline{c}^a \theta_a). \tag{4.78}$$

In order to calculate the effective action we now need to expand out the connections and curvatures order by order in γ_{ab}. This is done in appendix B. The action becomes

$$\int_{\mathcal{M}} d\mu \left(-\frac{1}{2\kappa^2} R + \frac{1}{2\kappa^2} G^{ab}\gamma_{ab} + \frac{1}{8\kappa^2} \gamma_{ab} \Delta_L^{(ab)(cd)} \gamma_{cd} + \frac{1}{\kappa} \overline{c}_a \Delta_G^{ab} c_b \right). \tag{4.79}$$

The Einstein tensor has been denoted here by $\mathbf{G}$. Explicit forms for the operators Δ_L and Δ_ϕ are given in appendix B. We can choose the simplest form of these operators (i.e. with gauge parameter $\alpha = 1$). The last term $\mathcal{L}_i$ containing all higher orders has been omitted.

The effective action at order $\hbar$ would be

$$\Gamma[\mathbf{g}] = I[\mathbf{g}] + \tfrac{1}{2}\hbar \log \det \Delta_L[\mathbf{g}] - \hbar \log \det \Delta_G[\mathbf{g}]. \tag{4.80}$$

Using ζ-function regularisation, the renormalisation scale enters the effective action with heat kernel coefficients from the graviton and the ghost,

$$\mu_R \frac{d\Gamma}{d\mu_R} = \frac{\hbar}{4\pi^2} \int_{\mathcal{M}} b_2(\mathbf{g}) \tag{4.81}$$

where $b_2(\mathbf{g}) = b_2(\Delta_L) - 2b_2(\Delta_G)$. These coefficients depend on squares of the Riemann tensor, terms that were not included in the original action.

Additional experiments would have to be performed to fix the coefficients of these curvature squared terms.

General expressions for the heat kernel coefficients are given in appendix A. Some simplification occurs because the classical field equations imply that the Ricci tensor vanishes. Therefore solutions to the quantum corrected equations have $R_{ab} = O(\hbar)$ at most. Terms in the effective action involving contractions of the Riemann tensor can only make a difference at order $\hbar^2$.

In four dimensions the Gauss–Bonnet identity relates Euler number to the Riemann tensor. The full expression is given in appendix D. This implies that

$$\mu_R \frac{d\Gamma}{d\mu_R} = \hbar \frac{212}{15} \chi. \tag{4.83}$$

The Euler number does not contribute to the field equations because it is a topological invariant and does not change under small variations of the metric. Therefore gravity without matter narrowly escapes being non-renormalisable at one-loop order.

Things change if we add matter fields. Including d massless scalars, as in appendix B, leads eventually to

$$\mu_R \frac{d\Gamma}{d\mu_R} = \hbar \frac{212 + d}{15} \chi + \hbar \frac{362 + d}{160\pi^2} \kappa^4 \int_{\mathcal{M}} d\mu (\nabla\phi \cdot \nabla\phi)^2. \tag{4.84}$$

The result has been simplified using the field equations $R_{ab} = \nabla_a\phi\nabla_b\phi + O(\hbar)$. The effective action is no longer renormalisable because of the last term.

5

Quantum statistical mechanics

Statistical mechanics is concerned with predicting the gross features of physical systems in terms of the microscopic dynamics. These features are parametrised by thermodynamic quantities like the temperature and will play an important role in some of the physical developments described later in this book.

A fundamental notion in statistical mechanics is an ensemble of different configurations that have the same thermodynamic variables. In quantum field theory we talk of an ensemble of quantum states. Each of these states is a many-body state.

The simplest systems in statistical physics are in thermodynamical equilibrium, and these are used for the definition of many of the thermodynamic variables. The systems with which we have to deal are mostly out of equilibrium but lie close to equilibrium over short time intervals. One of the aims of this chapter is to develop a description of how the thermodynamic variables evolve in time.

One of the most important thermal effects in the context of quantum field theory is the restoration of gauge symmetry. This was analysed in detail first by Weinberg (1974) and also Dolan and Jackiw (1974). This also raises the possibility of supercooling, and the resultant analysis of quantum tunnelling pioneered in field theory by Coleman (1977) and Callan and Coleman (1977), following similar ideas in non-relativistic statistical mechanics.

5.1 Finite-temperature field theory

The starting point for the discussion of thermal quantum fields is to assume that there is an ensemble of quantum states with a canonical distribution. This means that we fix the temperature and any conserved charges. The expectation value of an operator is given by thermal averaging of quantum expectation values,

$$\langle A \rangle = Z^{-1} \text{tr} \left(A \, e^{-\beta H} \right) \tag{5.1}$$

with the trace over quantum states. The inverse temperature $\beta = 1/T$ multiplies the Hamiltonian operator H. Normalisation introduces the partition function

$$Z(\beta, Q) = \text{tr}\left(e^{-\beta H}\right). \tag{5.2}$$

Thermodynamic variables can be obtained from the partition function and some are listed in table 5.1.

The operator $\exp(-\beta H)$ and the time evolution operator $\exp(iHt/\hbar)$ differ only by the replacement of $\beta = -it/\hbar$. We saw in chapter 2 that time evolution operators can be obtained from path integral expressions, and so we should expect a similar path integral expression for the partition function, but one which uses an imaginary time coordinate it (figure 5.1).

Consider, for example, a real scalar field. States can be labelled by the spatial field configuration $\widehat{\phi}$. The trace operation is made up of diagonal expectation values with identical configuration on the left and the right of the operator. The thermal average should therefore have a path integral formulation with fields which are periodic in the imaginary time coordinate, with period $\hbar\beta$. These fields are equivalent to fields on the manifold $M_3 \times S^1$, where M_3 is the spatial manifold.

Green functions can be obtained from a generating function $Z[J]$, evaluated on the cylinder,

$$Z_\beta[J] = \int d\mu[\phi]\, e^{-I_J/\hbar} \tag{5.3}$$

where

$$I_J = I - \int d\mu(x)J(x)\phi(x). \tag{5.4}$$

Functional differentiation with respect to the source and analytic continuation to real time will give thermally averaged expectation values. Now $\langle\phi(\mathbf{x})\phi(\mathbf{x}')\rangle$ is the thermal Green function and it plays a similar role to the Feynman Green function.

Functional differentiation of the effective action on the cylinder with respect to the metric gives the thermal average of the stress-energy tensor $\langle T_{ab}\rangle$. Comparing the vacuum amplitude with the partition function implies that

Table 5.1 Thermodynamic relations for the canonical ensemble.

Quantity	Symbol	Formula
free energy	F	$-\beta^{-1}\log Z$
energy	E	$-\partial_\beta \log Z$
entropy	S	$\beta\, \partial_\beta \beta^{-1}\log Z$

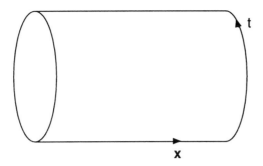

Figure 5.1   Periodic identification of the imaginary time coordinate.

$\Gamma_\beta = \hbar\beta F$ in thermal equilibrium, and some of the expressions for metric derivatives of the effective action are equivalent to familiar thermodynamic relations in table 5.1, such as $E = \partial(\beta F)/\partial\beta$.

Calculating the effective action perturbatively is very similar to the zero-temperature case, with order $\hbar$ contribution

$$\Gamma^{(1)} = \tfrac{1}{2}\hbar \operatorname{tr} \log \Delta. \tag{5.5}$$

In flat space, the functional trace is over functions on the cylinder. For a free scalar field of mass m ($\hbar$ has been absorbed into m),

$$\Gamma^{(1)} = \tfrac{1}{2}\hbar\Omega \sum_n \int \frac{d^3k}{(2\pi)^3} \log(k^2 + m^2 + (2\pi n/\hbar\beta)^2) \tag{5.6}$$

where $\hbar\boldsymbol{k}$ is the three-momentum eigenvalue, n is the S^1 mode number and Ω is the volume.

In the example below it is shown that the high-temperature limit of the free energy density becomes

$$\frac{F}{\Omega} \sim -\hbar^{-3}\frac{\pi^2}{90}T^4 + \hbar^{-1}\frac{1}{24}m^2T^2 - \frac{1}{12\pi}m^3T + V(\phi). \tag{5.7}$$

The energy density obtained from this (multiplied by 2 for the photon polarisation states) agrees with Planck's radiation law.

The partition function of vector boson fields can be calculated in the same way. The inclusion of conserved charges requires a small modification. The partition function for the canonical ensemble can be obtained from the grand canonical ensemble partition function,

$$Z(\beta,\mu) = \operatorname{tr}\left(e^{-\beta H - \mu Q}\right). \tag{5.8}$$

The time-like component of the vector boson already couples to the charge, due to the coupling in the Lagrangian between the vector bosons and the charged current, $i\mathbf{A}\cdot\mathbf{J}$. Therefore the partition function for the grand canonical ensemble can be obtained by including a vector gauge field $i\mu\mathbf{dt}$ in the path integral.

The partition function for Fermi fields has an important difference. A time displacement of $i\hbar\beta$ can be generated by two rotations through π radians. Fermi fields are therefore antiperiodic in the imaginary time direction.

Example: Derivation of the high-temperature expansion.

There are different ways to arrive at the high-temperature expansion. The one which follows uses the heat kernel representation of the ζ-function and the Poisson resummation formula. In general, it is possible to write the ζ-function as

$$\zeta(s) = \frac{1}{\Gamma(s)} \int_0^\infty dt\, t^{s-1} \sum_i e^{-\lambda_i t}. \tag{5.9}$$

At finite temperature,

$$\sum_i e^{-\lambda_i t} = \Omega \int \frac{d^3k}{(2\pi)^3} e^{-(k^2+m^2)t} \sum_n e^{-(2\pi n)^2 t/(\hbar\beta)^2}. \tag{5.10}$$

The sum over n can be rewritten in a more convenient way using the Poisson resummation formula (or a property of theta functions),

$$\sum_n e^{i\pi\tau n^2 + 2ni\pi v} = \sqrt{\frac{i}{\tau}} e^{-i\pi v^2/\tau} \sum_n e^{-i\pi n^2/\tau + 2ni\pi v/\tau}. \tag{5.11}$$

Also integrating over k,

$$\begin{aligned} \sum_i e^{-\lambda_i t} &= \frac{\hbar\Omega\beta}{(4\pi t)^{1/2}} \int \frac{d^3k}{(2\pi)^3} e^{-(k^2+m^2)t} \sum_n e^{-(\hbar\beta)^2 n^2/4t} \\ &= \frac{\hbar\Omega\beta}{16\pi^2 t^2} \sum_n e^{-(\hbar\beta)^2 n^2/4t - m^2 t}. \end{aligned} \tag{5.12}$$

The integral over t results in

$$\zeta(s) - \zeta_0(s) = \frac{\hbar\Omega\beta}{8\pi^2\Gamma(s)} {\sum_n}' \left(\frac{4m^2}{\hbar^2\beta^2 n^2} \right)^{1-s/2} K_{2-s}(\hbar\beta m n) \tag{5.13}$$

where $\zeta_0(s)$ is the zero-temperature result from the $n = 0$ term removed from the sum. At $s = 0$, the right-hand side vanishes because of the pole in $\Gamma(s)$.

Some useful identities at this point are

$$K_2(2\pi n z) = \frac{3z^2}{(2\pi n)^2} \int_0^\infty \frac{\cos 2\pi n t}{(t^2 + z^2)^{5/2}} dt \qquad (5.14)$$

and

$$B_4(\{t\}) = -\frac{3}{\pi^4} \sum_{n>0} \frac{\cos 2n\pi t}{n^4} \qquad (5.15)$$

where B_4 is a Bernoulli polynomial, and $\{t\}$ is the fractional part of t. These give,

$$\zeta'(0) - \zeta_0'(0) = \frac{\pi^2 \Omega z^4}{\hbar^3 \beta^3} \int_0^\infty dt \frac{\left(\frac{1}{30} - \{t\}^2 + 2\{t\}^3 - \{t\}^4\right)}{(t^2 + z^2)^{5/2}}. \qquad (5.16)$$

Replacing $\{t\}$ by t gives the first three terms in the expansion in powers of $z = \hbar \beta m / 2\pi$,

$$(\hbar \beta \Omega)^{-1} \zeta'(0) = \hbar^{-3} \tfrac{1}{45} \pi^2 T^4 - \hbar^{-1} \tfrac{1}{12} T^2 m^2 + \tfrac{1}{6} \pi^{-1} T m^3 + \dots . \qquad (5.17)$$

It is also possible to repeat the calculation with a chemical potential. The eigenvalues change by $2\pi n/\beta \to 2\pi n/\beta \pm i\mu$, and this results in the change $K_{2-s}(\hbar \beta m n) \to 2 K_{2-s}(\hbar \beta m n) \cosh(\beta \mu n)$ in $\zeta(s)$.

5.2   Finite-temperature effective potentials

The effective potential is defined as the leading term in a derivative expansion of the effective action,

$$\Gamma[\phi] = \int d\mu(x) \left(V_\beta(\phi) + Z(\phi)(\nabla \phi)^2 + \dots \right). \qquad (5.18)$$

In an interacting theory the minima of the effective potential give the free energies of distinct phases of the system, enabling phase transitions to be identified. The temperature-dependent corrections are particularly important if the scalar field is a Higgs field associated with symmetry breaking, because we have the possibility of different symmetry-breaking regimes at different temperatures.

For example in the $O(N)$ model with Higgs potential,

$$V(\Phi) = -\tfrac{1}{2} \mu^2 \Phi^2 + \tfrac{1}{4} \lambda \Phi^4. \qquad (5.19)$$

The effective action contains contributions from Higgs, vector and ghost fields,

$$\Gamma^{(1)} = \tfrac{1}{2} \hbar \log \det \Delta_A + \tfrac{1}{2} \hbar \log \det \Delta_\phi - \hbar \log \det \Delta_{gh} \qquad (5.20)$$

with masses given previously (section 4.4). We can display phase transitions on a graph of the effective potential $V_\beta(\phi)$ shifted vertically to pass through

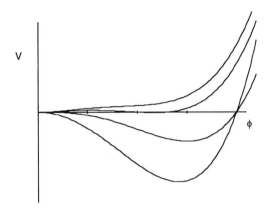

Figure 5.2   The finite-temperature effective potential for a range of
temperatures from $T = 0$ to T slightly larger than the critical temperature T_c.

the origin. Thus, if we keep only the vector boson masses $m_A^2 = \frac{1}{2}g^2\phi^2$
($m_A^2 = \frac{1}{4}g^2\phi^2$ for SU(N)),

$$V_\beta(\phi) = \tfrac{1}{8}\hbar^{-1}(N-1)m_A^2 T^2 - \tfrac{1}{4}\pi^{-1}(N-1)m_A^3 T + V(\phi) \qquad (5.21)$$

shown in figure 5.2. At high temperatures the minimum of the potential lies at
the point of symmetry $\phi = 0$. As the temperature decreases a new minimum
appears. The two minima have the same energy at a critical temperature T_c.

The real situation is rather more complicated than this picture would
suggest due to problems with truncating the perturbation series at order $\hbar$. At
high temperatures there are two expansion parameters, $\hbar$ and β. Close to the
critical temperature the effective mass is a third small parameter. Orders of
perturbation theory in $\hbar$ can mix. This is manageable at high temperatures,
but particularly difficult at temperatures close to the critical temperature.

5.3   Quantum tunnelling

So far distinct phases have been associated with minima of the effective
potential. Some of these minima may be higher than the global minimum
of the potential and these may be unstable to quantum or thermal energy
fluctuations. A metastable phase can sometimes be established at the local
minimum if the tunnelling rate is very slow compared to other physical
processes.

This should be familiar from everyday experience. For example, water

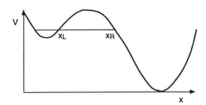

Figure 5.3 Tunnelling from a local minimum in the potential.

vapour cooled below the temperature at which it condenses can suddenly form droplets of water. The potential barrier in this case is the extra energy needed to form the surface of a water droplet.

Tunnelling rates can be calculated very efficiently by using path integral techniques. It will be helpful first to examine a simple zero-temperature system with one degree of freedom, a particle with position x in a local potential behind a barrier $V(x)$, as shown in figure 5.3.

The classical solutions satisfy

$$\ddot{x} + V'(x) = 0. \tag{5.22}$$

For energy E less than the barrier height, the particle bounces at x_1 where $\dot{x} = 0$. The amplitude for $x > x_1$ can be obtained by solving the imaginary-time problem

$$-\ddot{x} + V'(x) = 0 \tag{5.23}$$

for the 'bounce' solution, i.e. $\dot{x} = 0$ at $x = x_1$.

What is really going on here? Analytic continuation often arises in contour integration, but the time variable does not appear as a variable of integration in the path integral. We can rectify this by introducing an intrinsic time parameter τ, which also means introducing a lapse function N and energy function E. The action, for fixed values of x and t at the endpoints of a path, is then

$$S[x, t, N, E] = \int_0^1 \left(-iE + N(\tfrac{1}{2}N^{-2}\dot{x}^2 - V(x) + E)\right) d\tau. \tag{5.24}$$

The path integral measure is now over the set of fields x, t, N, E,

$$\langle x_1, T | x_0, 0 \rangle = \int_0^1 d\mu[x, t, N, E] \mathcal{P}(x_1, t_1 | x_0, t_0) e^{iS[x,t,N,E]/\hbar} \tag{5.25}$$

where the projection operator reminds us that the endpoints are fixed.

Let the initial wavefunction of the particle be $\psi(x) = \exp(-\frac{1}{2}\omega x^2)$, where $\omega^2 = V''(0)$. The probability of finding the particle in the same state at a later time is

$$\langle \psi, T | \psi, 0 \rangle = \int d\mu(x_0, x_1)\, \psi(x_1)\psi(x_0)\langle x_1, T | x_0, 0 \rangle. \tag{5.26}$$

We approximate this path integral by saddle point paths. Variation of N gives the equation

$$\tfrac{1}{2}N^{-2}\dot{x}^2 + V(x) = E \tag{5.27}$$

and it is possible to choose a gauge in which N is constant.

If the energy $E(x_0, x_1, T)$ is less than the barrier height, then there is a solution which oscillates $\omega T/2\pi$ times between x_0 and x_1. This solution is the same one that would be obtained for a simple quadratic potential.

It is also possible to obtain other solutions by moving the variable N onto the complex plane. Solutions with a piecewise constant N, such that $N \propto i$ under the barrier, represent quantum tunnelling. These paths $x(\tau)$ are saddle points of the path integral provided $\dot{x} = 0$ at the bounce points x_L or x_R where $E - V$ changes sign. The part of the solution from x_L to x_R is a bounce solution x_b.

The bounce can happen on any one of the oscillations. The amplitude to remain in the same state is therefore given by a sum of contributions from each path,

$$\langle \psi, T | \psi, 0 \rangle = \langle \psi, T | \psi, 0 \rangle_{\text{oscillator}} \left(1 + i\mathcal{F}e^{-B} + \ldots \right). \tag{5.28}$$

The factor of i in the second term comes from the lapse function N. The exponent $\hbar B = -iS[x_b] - iET$,

$$\hbar B = -i \int_0^1 \left(\tfrac{1}{2}N^{-2}\dot{x}^2 - V(x) + E \right) N d\tau. \tag{5.29}$$

Using equation (5.27) and $N = i$ we can rewrite this as

$$B = \frac{1}{\hbar} \int_{x_L}^{x_R} \left(2(V - E) \right)^{1/2} dx. \tag{5.30}$$

This is, of course, familiar from the WKB (Wentzel–Kramers–Brillouin) approximation to solutions of the Schrödinger equation.

The factor $\mathcal{F}$ is a contribution from the fluctuations about the path. Expanding the action to second order gives the fluctuation operator,

$$\Delta = -\nabla^2 + V''(x_b). \tag{5.31}$$

by differentiating the field equation,

$$(-\nabla^2 + V''(x_b))\dot{x}_b = 0. \tag{5.32}$$

The function $\dot{x}_b$ is therefore very close to being an eigenfunction with a zero eigenvalue, and has to be excluded from the range of the operator. This exclusion will be denoted by a prime. The mode corresponds physically to translating the bounce solution in the time direction. Its contribution to the path integral can be evaluated separately; it will give the total time T multiplied by a factor for normalisation of the mode function.

The function $\dot{x}_b$ has one node and the theory of differential equations predicts that there is another mode function with a negative eigenvalue, leading to a factor of i in the square root of the determinant. Therefore $\mathcal{F} = \frac{1}{2}i|A|T$ where

$$|A| = \left| \frac{\det \Delta'}{\det \Delta_0} \right|^{-1/2} \left(\frac{B}{2\pi} \right)^{1/2}. \tag{5.33}$$

The probability to remain in the state ψ is now

$$|\langle \psi, T | \psi, 0 \rangle|^2 = 1 - T|A|e^{-B}. \tag{5.34}$$

Alternatively, the rate of escape N is

$$N = |A| \exp(-B). \tag{5.35}$$

This result could be obtained more easily from the WKB approximation to the Schrödinger equation. The advantages of the path integral approach only become compelling when we have to consider relativistic field theory.

5.4 Quantum tunnelling in field theory

The tunnelling rate from the metastable state in quantum field theory follows simply by replacing x with the full degrees of freedom of the quantum field,

$$N = |A| \exp(-B). \tag{5.36}$$

The exponent B is the difference in action between the bounce solution and the metastable state.

The bounce solution is now any field configuration on the four-dimensional Riemannian space M_4 or $M_3 \times S_1$ that satisfies the field equations and asymptotically approaches the metastable value. Tunnelling will be predominantly associated with the solution that has the smallest action simply because it gives the largest rate. This is usually the solution with the most symmetry. In flat space the simplest bounce solution would be the one with

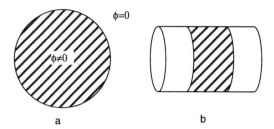

Figure 5.4   The bounce solution surrounded by the old phase at zero temperature (a) and non-zero temperature (b).

Figure 5.5   Real time evolution of the $O(4)$ symmetric bubble.

because it gives the largest rate. This is usually the solution with the most symmetry. In flat space the simplest bounce solution would be the one with $O(4)$ symmetry and $\phi = \phi_b(|\mathbf{x}|)$. If the temperature is very high we might have to consider the solution with $O(3)$ symmetry instead, shown in figure 5.4. These solutions are not minima of the action because the action can be reduced in the direction of the mode with the negative eigenvalues.

The tunnelling rate gives the number of quantum transitions per unit volume and per unit time. Each nucleation event produces a region of the new phase in the form of a slice of the bounce solution, which we call a bubble of the new phase. There are four spacetime directions in which the $O(4)$ bounce solution can be displaced, and four zero modes to the fluctuation operator. The prefactor is now

$$|A| = \left|\frac{\det \Delta'}{\det \Delta_0}\right|^{-1/2} \left(\frac{B}{2\pi\hbar}\right)^2 . \tag{5.37}$$

After the bubble has nucleated we can follow its evolution by using the effective field equations. At zero temperature the time evolution gives the same result as analytic continuation of the bounce solution $\phi = \phi_b(|\mathbf{x}|^2 - t^2)$.

same result as analytic continuation of the bounce solution $\phi = \phi_b(|x|^2 - t^2)$. The wall of the bubble follows a hyperbolic trajectory of constant proper acceleration shown in figure 5.5, until this bubble collides with another bubble solution.

5.5 Phase transitions

The hot early stages of the universe would have been the likely setting for many phase transitions. The standard model of particle physics gives a clue as to the energies involved. The particle content of the standard model is shown in table ??. These particles are grouped into three families, the mass increasing as we move along. Within each family, particles are related by gauge symmetry, broken in the case of the electroweak fields.

Table 5.2   Particle content of the standard model.

Type	Particle
leptons	$\binom{\nu_e}{e}$, $\binom{\nu_\mu}{\mu}$, $\binom{\nu_\tau}{\tau}$
quarks	$\binom{u}{d}$, $\binom{c}{s}$, $\binom{t}{b}$
gluons	$g_1 \ldots g_8$
vectors	$W^\pm$, Z, photon
scalars	Higgs

One or more phase transitions would be expected to occur in connection with a change in phase from free quarks to nucleons. The quarks pick up masses around the same time due to their interactions with coherent states of gluons.

The unification of electromagnetism and the weak force would also be expected to lead to a phase transition in the early universe. This theory has two coupling constants g and g'. In the low-temperature phase a component ϕ of the Higgs field fixes the mass of three gauge bosons, the $W^\pm$ bosons each with mass $m_W = \frac{1}{2}g\phi$ and the Z boson with mass $\frac{1}{2}(g^2 + g'^2)^{1/2}\phi$. The finite-temperature effective potential would be similar to the one we saw earlier (equation (5.21)), substituting the new electroweak vector boson masses.

It is important to find out whether the Higgs field might have been trapped in the symmetric phase as the universe cooled down. If this happened, then quantum tunnnelling to the new phase would cause a very uneven density distribution and thermodynamic equilibrium would break down. This loss of

Table 5.3 Thermal history of the universe near the electroweak/quark-hadron phase transitions.

Temperature	g_s	horizon (kg)	$\log_{10}(1+z)$
1 MeV	11	3.6×10^{34}	9.6
100 MeV	11	3.6×10^{30}	11.6
1 GeV	74	1.4×10^{28}	12.6
100 GeV	97	1.2×10^{24}	14.6
1 TeV	106	1.2×10^{22}	15.6
100TeV	106	1.2×10^{18}	17.6

the universe today, such as the baryon density or the helium abundance. The density variation itself has a limited physical range set by the velocity of light to a distance ct. This is often called the distance to the physical horizon and the mass of radiation enclosed by the horizon is given in table 5.5.

According to the effective potential there is a small potential barrier at the critical temperature, but the perturbative expansion used to calculate it becomes suspect. If we use the approximation regardless, then we need to compare the tunnelling rate with the expansion rate of the universe. If the tunnelling rate is smaller then the phase transition is supercooled. The barrier decreases in height as the temperature falls. The phase transition will complete when the tunnelling rate drops to the same value as the expansion rate.

The expansion rate H is related to the radiation density ρ through $H^2 = \frac{8}{3}\pi G\rho$, where

$$\rho = \frac{1}{30}\pi^2 g_s \hbar^{-3} T^4. \tag{5.38}$$

Values of the degeneracy factor g_s are given in table 5.3.

What actually happens cannot be fully determined without knowing the Higgs mass. The current limits imply that there could only be a very small amount of supercooling. It is, of course, also possible that new physical effects will be present at the electroweak energy scales. Only experiment can resolve these issues.

6

Classical gravity

The theory of general relativity developed by Albert Einstein has an unparallelled combination of mathematical elegance and range of physical applicability. We know now from theorems of Stephen Hawking and Roger Penrose that there are situations where the framework of general relativity breaks down. The limitations of the theory are reached at the endpoints of gravitational collapse and at the time when the universe started expanding, where there are singularities in the curvature of spacetime.

Some exact solutions to the classical equations of general relativity are featured in this chapter, especially those describing simple versions of gravitational collapse or simple cosmological models. More detailed accounts of these solutions can be found in textbooks on general relativity, such as Misner et al. (1973) or Hawking and Ellis (1973).

6.1 Gravitational collapse

The largest objects in the cosmos are under constant threat of collapsing due to the force of gravity. Stars support themselves by the constant expenditure of energy and when the fuel burns out they have to contract to small white dwarfs or neutron stars. If their final mass is larger than the bound discovered by Chandrasekhar then there is no known force that can prevent their collapse. The Chandrasekhar limit for white dwarfs is set by the equilibrium of a degenerate electron gas at around 1.2 solar masses. For neutron stars the limit depends a little on the equation of state for nuclear matter, but the best available estimates come out around 2.4 solar masses.

A very compact object can trap light rays within its gravitational field and become a black hole. The spacetime boundary of trapped light is the event horizon of the black hole. It is a boundary but not a barrier to inward-travelling particles. In S.I. units the event horizon of a spherical black hole of mass M has a radius of $r = 2GM/c^2$. For a very massive cloud of gas, the pressure and density of the gas as the cloud crosses its own event horizon can be moderate and can fall well within the range of values where the physical laws are well understood. It seems, therefore, that in principle the formation of black holes cannot be avoided by a resort to 'unknown physics'.

Solving the Einstein equations for situations with pressure or dissipation and non-spherical geometry is a difficult problem, requiring numerical methods and powerful computers. Nevertheless, much useful information can be obtained from analytical arguments, especially the global techniques first developed by Hawking and Penrose. The most famous results are the singularity theorems that can be used to indicate when singularities will form. These theorems are based upon the idea that when a bunch of world-lines are converging together then matter flowing along the lines becomes focused and the curvature rises, causing more convergence. If the lines converge into a caustic, then a singularity forms (not necessarily at the caustic). A singularity is said to form when it becomes impossible to continue spacetime indefinitely along all of the world-lines.

This is particularly likely to happen if there exists a surface called a trapped surface. A two-dimensional surface splits up the tangent space of the embedding spacetime into two tangential and two normal directions. Usually, the normals in one direction have positive divergence and in the other negative. The surface is trapped when both normals have negative divergence.

An important singularity theorem goes as follows:

Theorem A spacetime is singular if:

1. $R_{ab}K^aK^b \geq 0$ for all non space-like vectors $\mathbf{K}$.
2. $K_{[a}R_{b]cd[e}K_{f]}K^cK^d \neq 0$ for a vector $\mathbf{K}$ tangent to some geodesic.
3. There are no closed time-like curves.
4. Either:
   (a) there is a closed trapped surface; or
   (b) there is a point p for which $\nabla \cdot \mathbf{K} < 0$ for all of the vectors $\mathbf{K}$ tangent to the past light cone of p.

The first requirement can be viewed as a condition on the stress–energy tensor, because due to the Einstein equations,

$$R_{ab}K^aK^b = 8\pi G(T_{ab} - \tfrac{1}{2}Tg_{ab})K^aK^b. \tag{6.1}$$

If the right-hand side of this equation is positive for all non space-like vectors then we say that the stress tensor satisfies the strong energy condition. If this holds for null vectors then the stress tensor satisfies the dominant energy condition.

Item 2 is needed only to exclude some highly atypical situations. The third condition is more interesting. It is often called the chronology condition, because it would be impossible to order events in time without it. This condition is not, as far as we know, implied by the Einstein equations.

The fourth condition depends on the particular application. In gravitational collapse the more likely version would be 4(a), whilst for cosmology and the Big Bang the more useful condition would be 4(b).

Beside the existence of singularities, there are other important issues arising from gravitational collapse. The eventual form of the spacetime outside of the collapsing body raises several important questions including: 'Does the spacetime settle into a stationary state?', and 'Are the singularities hidden by an event horizon?'.

The first question is about asymptotic behaviour in time and can be partially examined by perturbing a purely spherical collapse. The answer seems to be affirmative (Price 1972). Asymmetric modes are radiated away as gravitational waves. Only two modes survive, the constant spherically symmetric mode and one constant rotational mode.

The answer to the second question, whether singularities are hidden by event horizons, would be affirmative according to the cosmic censorship hypothesis (Penrose). This is actually a weak form of cosmic censorship. There is also a stronger version stating that singularities are invisible to all observers, not just those outside of the event horizon. Any proof of the hypothesis in its strong or weak form would certainly involve conditions on the matter content and probably also on the initial conditions. What constitutes reasonable conditions is hotly disputed and the hypothesis is presently under attack.

6.2 Collapsing dust

The case of a collapsing spherical dust cloud can be followed using analytical methods. The corresponding solution of the field equations is originally due to Oppenheimer and Schneider or in general form to Tolman and Bondi. This solution begins with the spherically symmetric metric,

$$ds^2 = -e^{2\Psi} dt^2 + e^{2\Lambda} dr^2 + R^2(d\theta^2 + \sin^2\theta \, d\phi^2) \qquad (6.2)$$

where $\Psi(r,t)$, $\Lambda(r,t)$ and $R(r,t)$ are the only free functions. The relevant components of the Einstein tensor in an orthonormal frame are given in table 6.2, where $\widehat{\partial}_t$ denotes $e^{-\Psi}\partial/\partial t$ and $\widehat{\partial}_r$ denotes $e^{-\Lambda}\partial/\partial r$.

Some simplification occurs if we use the proper time as a time coordinate and take $\Psi = 1$. It is also useful to introduce new functions $f(r,t)$ and $m(r,t)$ defined by

$$f = 1 - e^{-2\Lambda} R'^2 \qquad (6.3)$$

and also

$$Gm = \tfrac{1}{2} R(\dot{R}^2 - f). \qquad (6.4)$$

Table 6.1 Some components of the Einstein tensor for the spherical metric element.

G	Expression
$G^{\hat{r}}_{\hat{t}}$	$2R^{-1}\hat{\partial}_t R \hat{\partial}_r \Psi - 2R^{-1}\hat{\partial}_t \hat{\partial}_r R$
$G^{\hat{r}}_{\hat{r}}$	$-2R^{-1}\hat{\partial}_t \hat{\partial}_t R + 2R^{-1}\hat{\partial}_r R \hat{\partial}_r \Psi + R^{-2}\left((\hat{\partial}_r R)^2 - (\hat{\partial}_t R)^2 - 1\right)$
$G^{\hat{t}}_{\hat{t}}$	$2R^{-1}\hat{\partial}_r \hat{\partial}_r R - 2R^{-1}\hat{\partial}_r R \hat{\partial}_r \Lambda + R^{-2}\left((\hat{\partial}_r R)^2 - (\hat{\partial}_t R)^2 - 1\right)$

The Einstein equations relate $\dot{f}$ and $\dot{m}$ to the density $\rho = T_{tt}$, pressure $p = T^r_r$ and momentum flux $s = T_{rt}$,

$$\dot{f} = 8\pi GR(R')^{-1}(1-f)\,s \qquad (6.5)$$

$$\dot{m} = 4\pi R^2(R')^{-1}(1-f)\,s - 4\pi R^2 \dot{R}\,p \qquad (6.6)$$

$$m' = 4\pi R^2 R'\rho - 4\pi R^2 \dot{R}\,s \qquad (6.7)$$

For dust, where only ρ is non-zero, both m and f become functions of r alone. The remaining Einstein equation for m' has the interpretation that $m(r)$ measures the mass within a radius r. After m and f are determined from the initial data $R(r, 0)$, $\dot{R}(r, 0)$ and $\rho(r, 0)$, we can use equation (6.4) in the form

$$\dot{R}^2 + V(R, r) = 0 \qquad (6.8)$$

to determine $R(r, t)$.

Parametric solutions can easily be found for $R(r, t)$ when $\dot{R}(r, 0) = 0$,

$$R = \tfrac{1}{2}r(1 + \cos\eta), \qquad t = \left(\frac{r^3}{8Gm}\right)^{1/2}(\eta + \sin\eta). \qquad (6.9)$$

The density and curvature are singular at $\eta = \pi$, where the matter collapses to a point. The collapsing cloud is drawn in figure 6.1. On this figure the outgoing light rays from the centre of the cloud are also shown. The event horizon is generated by the light rays which approach $R = 2GM$, where M is the total mass of the cloud.

6.3 Black holes

All of the indications are that the final result of gravitational collapse, however complicated the system during the early stages, will be a stationary black hole. In this final form it is possible to show further that the surface of trapped

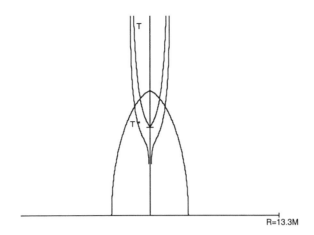

Figure 6.1   The collapse of a spherical cloud in (R,t) coordinates. The surface of
the star is at constant r. Also shown are the event horizon and the surface
$$R = 2GM.$$

light has a beautifully simple geometry. Residual properties that characterise
the black hole exterior following an electrically neutral collapse are only its
mass and rotation.

The fully symmetrical case of a black hole in empty space, with no electric
charge or angular momentum, results in the Schwarzschild metric. This can be
found by equating the Einstein tensor to zero. Taking $\Psi(r)$, $\Lambda(r)$ and $R = r$
gives the metric in Schwarzschild coordinates,

$$ds^2 = -\left(1 - (2GM/r)\right) dt^2 + \left(1 - (2GM/r)\right)^{-1} dr^2 + r^2(d\theta^2 + \sin^2\theta \, d\phi^2).$$
$$(6.10)$$

It is not even necessary to assume that the metric components are independent
of time because the assumptions of spherical symmetry and vacuum are
sufficient to derive the metric.

The event horizon is at $r_h = 2GM$, where the radial metric component
diverges. In order to verify that this is the event horizon it is necessary to
consider how light moves through the spacetime. It is also advantageous to
set up a coordinate system in which the metric is regular at the event horizon.

Null geodesics in the purely radial direction satisfy $\dot{t} = E$ and $\dot{r}^* = \pm E$,
where r^* is a modified radial coordinate $r^* = r + r_h \log(r/r_h - 1)$. As $r \to r_h$,
the coordinate $r^* \to -\infty$. The Kruskal coordinates,

$$U = -\exp(-\kappa_h(t - r^*)) \quad V = \exp(\kappa_h(t + r^*)) \tag{6.11}$$

where κ_h is a constant, are constant along the radial light rays and finite on the horizon. In these coordinates

$$ds^2 = -\kappa_h^{-2}(1 - r_h/r)^{1-2\kappa_h r_h} e^{-2\kappa_h r} dU\, dV + r^2(d\theta^2 + \sin^2\theta\, d\phi^2). \quad (6.10)$$

The metric is regular at the horizon for $\kappa_h = 1/2r_h$. This value is known as the surface gravity of the black hole.

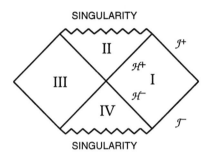

Figure 6.2   A Penrose diagram of Schwarzschild.

It is especially helpful for understanding the global properties of the spacetime to arrange for the entire spacetime to be covered by finite coordinate regions; then slices of the spacetime can be drawn in full on the page. If radial light rays are parallel lines in the new coordinates, then the resulting diagram is called a Penrose diagram. A Penrose diagram for the extended black hole spacetime is shown in figure 6.2. Hyperbolic tangents of the Kruskal coordinates have been used to bring the regions far from the hole into the diagram.

The original coordinates only covered the region labelled I in the figure. The asymptotic boundaries of region I, as $r \to \infty$, are labelled $\mathcal{J}^+$ and $\mathcal{J}^-$, 'scri plus' and 'scri minus'. The other two boundaries include the event horizon $\mathcal{H}^+$, marking the boundary of the past of $\mathcal{J}^+$.

The region labelled III is similar to I but it uses new r and t coordinates. These can be introduced in terms of the Kruskal coordinates by the replacement

$$U = \exp(-\kappa_h(t - r^*)) \quad V = -\exp(\kappa_h(t + r^*)). \quad (6.11)$$

Both sets of asymptotic regions are connected by space-like sections through the spacetime. Horizontal sections form the Einstein–Rosen bridge between the two regions. However, from the diagram it should be clear that no observer, who is confined by causality, can cross from one side to the other.

6.4 Other types of black holes

A black hole with electric charge is a solution of the Einstein equations with electromagnetic source terms. The metric for a charged black hole has the Reissner–Nordstrom form,

$$ds^2 = -\frac{\Delta}{r^2}dt^2 + \frac{r^2}{\Delta}dr^2 + r^2(d\theta^2 + \sin^2\theta\, d\phi^2) \qquad (6.12)$$

where

$$\Delta = r^2 - 2GMr + (Q/4\pi)^2. \qquad (6.13)$$

The electric field is described by a vector potential $\mathbf{A} = (Q/4\pi r)\mathbf{dt}$. It represents a radial field with charge Q.

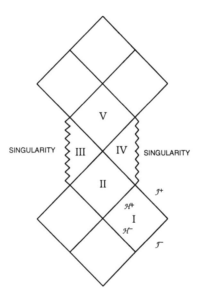

Figure 6.3 Part of the Penrose diagram of Reissner–Nordstrom. The diagram extends vertically.

This metric has two horizons at the roots $r_\pm$ of $\Delta = 0$, an outer horizon at r_+ and an inner horizon at r_-. Kruskal coordinates can be defined for the charged black hole as before, but now regularity of the coordinates on the horizon at r_h requires

$$\kappa_h = |\Delta'(r_h)|/2r_h^2 = (r_+ - r_-)/2r_h^2. \qquad (6.14)$$

New sets of Kruskal coordinates can be defined for each horizon in turn, building up the diagram shown in figure 6.3.

The outer horizons are event horizons for the part of the universe in which they occur. The inner horizons introduce a new feature, namely that an extra piece of the singularity becomes visible each time the inner horizon is crossed, as shown in figure 6.3.

The inner horizon can be described in terms of causal properties of the spacetime, for which it is worth introducing some terminology. Infinite spatial surfaces in the spacetime are called Cauchy surfaces. Points through which every time-like curve would intersect a Cauchy surface make up the domain of dependence of that Cauchy surface. Solutions to the wave equation can be constructed within the domain of dependence of their Cauchy data. The boundary of the domain of dependence is the Cauchy horizon. The inner horizon of the charged black hole is a Cauchy horizon.

Perturbations of the Reissner–Nordstrom spacetime have been analysed extensively and the evidence suggests strongly that a singularity forms on the inner horizon. Therefore for real physical observers, crossing the inner horizon and catching a view of the singularity seems to be impossible. In other words, the cosmic censorship conjecture seems to apply if the spherically symmetric case is excluded.

Spherical symmetry could arise from the collapse of a non-rotating body, but the angular momentum of a typical collapsing body would result in a rotating black hole (figure 6.4). The relevant metric is the Kerr metric,

$$ds^2 = \rho^2(\Delta_r^{-1}dr^2 + \Delta_\mu^{-1}d\mu^2) + \rho^{-2}\Delta_\mu \, \boldsymbol{\omega}^\phi \otimes \boldsymbol{\omega}^\phi - \rho^{-2}\Delta \, \boldsymbol{\omega}^t \otimes \boldsymbol{\omega}^t \qquad (6.15)$$

with 1-forms,

$$\boldsymbol{\omega}^\phi = \mathbf{dt} - a^{-1}\sigma_r^2 \mathbf{d\phi} \qquad \boldsymbol{\omega}^t = \mathbf{dt} - a^{-1}\sigma_\mu^2 \mathbf{d\phi}. \qquad (6.16)$$

The parameters in the metric are the mass M and the rotation parameter a,

$$\sigma_r^2 = a^2 + r^2 \qquad \Delta_r = r^2 - 2GMr + a^2 \qquad (6.17)$$

$$\sigma_\mu^2 = \Delta_\mu = a^2 - \mu^2 \qquad \rho^2 = r^2 + \mu^2. \qquad (6.18)$$

This metric has coordinate singularities at the roots of $\Delta_\mu = 0$ and $\Delta_r = 0$. If we identify $\mu = a\cos\theta$, then the first set of coordinate singularities are north and south polar singularities on surfaces of constant r and t.

For $a^2 < G^2M^2$, there are two roots $r_\pm$ of $\Delta_r = 0$. The larger root r_+ gives the location of the event horizon, as we shall see below, and r_- is an inner or Cauchy horizon.

Coordinate systems can be set up which make the metric regular on any particular horizon. In the region outside the event horizon there exist two simple families of null geodesics with tangent vectors

$$\dot{t} = \frac{\sigma_r^2}{\Delta_r}E \qquad \dot{r} = \pm E \qquad \dot{\theta} = 0 \qquad \dot{\phi} = \frac{a}{\Delta_r}E \qquad (6.19)$$

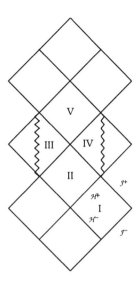

Figure 6.4   Part of the Penrose diagram of a rotating hole.

where a dot denotes derivative with respect to geodesic affine parameter and
E is a constant. Corresponding 'light-like' coordinates which are fixed along
each family can therefore be defined by $u = t - r^*$ and $v = t + r^*$, where
$dr^* = \Delta_r^{-1}\sigma_r^2 dr$.

There is a unique combination of the Killing vectors that is null on the
(future) event horizon, $\boldsymbol{l} = \mathbf{e}_t + \Omega_+ \mathbf{e}_\phi$, where $\Omega_+ = a\sigma_r^{-2}(r_+)$. The surface
gravity κ is defined in terms of $\boldsymbol{l}$, through $\nabla(\boldsymbol{l}^2) = \kappa \boldsymbol{l}$, which implies

$$\kappa_h = |\Delta_r'(r_h)|/2\sigma_r(r_h). \tag{6.20}$$

A new angular variable $\phi_+ = \phi - \Omega_+ t$, which is constant along the integral
curves of $\boldsymbol{l}$, should be used in place of ϕ. Along the null geodesic congruence,
ϕ_+ approaches a constant as the horizon is approached but ϕ diverges.

In general relativity, rotating bodies can influence inertial frames and cause
them to move in the direction of rotation. This is strikingly shown by the
behaviour of the Killing field $\mathbf{e}_t$ along the time direction, which becomes
space-like before reaching the event horizon, in a region called the ergosphere.
The boundary of the ergosphere is at

$$r^2 - 2GMr + a^2 \cos^2 \theta = 0. \tag{6.21}$$

Inside the ergosphere it is impossible for an observer to be at rest.

6.5   de Sitter space

We turn now from the description of gravitationl collapse to models of the expanding universe. Historically the first cosmological model from the era of general relativity was the de Sitter model. This model resurfaced as the geometry behind the steady–state universe but its modern application is the basis for the inflationary universe, which we will consider in detail later.

The de Sitter spacetime is very special because it is one of only three four-dimensional spacetimes with the maximum amount of symmetry. This symmetry tells us that the metric can be obtained by embedding the hyperboloid

$$\alpha^2 = X^2 + Y^2 + Z^2 + W^2 - V^2 \qquad (6.22)$$

in five-dimensional Minkowski space,

$$ds^2 = dX^2 + dY^2 + dZ^2 + dW^2 - dV^2. \qquad (6.23)$$

de Sitter space is also a solution to the Einstein field equations with a source term $T_{ab} = -\Lambda g_{ab}$, where Λ is the cosmological constant. The radius α is related to the cosmological constant by $\alpha^2 = 3/\Lambda$.

Figure 6.5   Embeddings of de Sitter space with (a) horizontal time slicing, (b) oblique time slicing and (c) confocal time slicing.

The de Sitter metric comes in differing forms depending on how we choose the surfaces of constant time. Three choices are tabulated in table 6.2 and drawn in figure 6.5. Angular coordinates are described there in terms of a unit vector $\mathbf{n} = (\sin\theta\cos\phi, \sin\theta\sin\phi, \cos\theta)$, and spatial coordinates by $\mathbf{x} = (x, y, z)$. In all but the first coordinate system the coordinates only cover part of the whole spacetime.

The first three examples result in homogeneous space-like sections (table 6.3), but in one the sections are closed and have positive curvature whilst in the others they are open. In the open cases the coordinates only cover part of

Table 6.2 Coordinates for the embedding of de Sitter space.

Slicing	V	W	$\mathbf{X}$
horizontal	$\alpha \sinh(t/\alpha)$	$\alpha \cosh(t/\alpha) \cos \chi$	$\alpha \mathbf{n} \cosh(t/\alpha) \sin \chi$
oblique	\multicolumn{2}{c}{$((\alpha^2 \pm r^2)e^{t/\alpha} - \alpha^2 e^{-t/\alpha})/2\alpha$}	$\mathbf{x} e^{t/\alpha}$
vertical	$\alpha \cos \chi \sinh(t/\alpha)$	$\alpha \cosh(t/\alpha)$	$\alpha \mathbf{n} \sinh(t/\alpha) \sin \chi$
confocal	$\sqrt{\alpha^2 - r^2} \sinh(t/\alpha)$	$\sqrt{\alpha^2 - r^2} \cosh(t/\alpha)$	$r\mathbf{n}$

Table 6.3 de Sitter metrics in different coordinate systems.

Slicing	metric
horizontal	$-dt^2 + \alpha^2 \cosh^2(t/\alpha) \left(d\chi^2 + \sin^2 \chi(d\theta^2 + \sin^2 \theta \, d\phi^2)\right)$
oblique	$-dt^2 + \exp(2t/\alpha)(dx^2 + dy^2 + dz^2)$
vertical	$-dt^2 + \alpha^2 \sinh^2(t/\alpha) \left(d\chi^2 + \sinh^2 \chi(d\theta^2 + \sin^2 \theta \, d\phi^2)\right)$
confocal	$-(1 - r^2/\alpha^2)dt^2 + (1 - r^2/\alpha^2)^{-1}dr^2 + r^2(d\theta^2 + \sin^2 \theta \, d\phi^2)$

the spacetime and there are extentable geodesics that reach back to $t \to -\infty$ in only a finite proper time. This can be deduced from the Penrose diagram shown in figure 6.6.

The spatially flat coordinate system also has a conformally flat form obtained by replacing t with $\eta = -\alpha \exp(t/\alpha)$, where $-\infty < \eta < 0$. The last coordinate choice in the table gives a metric much closer in appearance to a black hole metric. The coordinates cover only one region of the conformal diagram shown in figure 6.6, with cosmological horizons at $r_h = \alpha$ which play the same role as the black hole event horizons. The surface gravity of these horizons is $\kappa_h = 1/\alpha$.

6.6 Naked singularities

All of the black hole metrics considered earlier become flat far away from the hole and therefore share the asymptotic structure of flat space locally. There exists a simple generalisation of these metrics which approach constant non–zero curvature asymptotically, at the price of introducing a cosmological

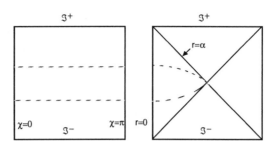

Figure 6.6   A Penrose diagram of de Sitter space

constant. The metric is the same as before (equation (6.15)), with

$$\Delta_r = (a^2 + r^2)(1 - \tfrac{1}{3}\Lambda r^2) - 2GMr + Q^2 \qquad (6.24)$$

$$\Delta_\mu = (a^2 - \mu^2)(1 + \tfrac{1}{3}\Lambda\mu^2). \qquad (6.25)$$

This spacetime can posess as many as three horizons. The parameter space is plotted in figure 6.7.

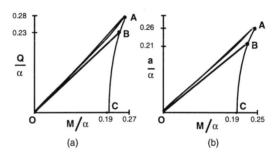

Figure 6.7   The parameter space of (a) charged and (b) rotating holes in de Sitter space. Coincident horizons occur on the outer boundaries of the triangular regions, $r_2 = r_3$ on OA and $r_2 = r_3$ on AC.

There are many asymptotic regions where the spacetime geometry is similar to de Sitter (figure 6.8). In the same way as in Reissner–Nordstrom, there are world-lines that pass through the black hole from one universe (i.e. asymptotic region) to another. Inside the black holes are naked singularities which violate the cosmic censorship hypothesis.

In the Reissner–Nordstrom case the global structure is unstable and the tiniest change, due to the presence of an observer for example, restores the effects of cosmic censorship. This instability is caused by radiation entering a black hole, which becomes infinitely blue-shifted in the vicinity of the Cauchy horizon, resulting in a diverging energy density. The situation in de Sitter space is different, as light rays running along the Cauchy horizon have travelled from a cosmological horizon and not from infinity. The difference in surface gravities of the two horizons determines whether the wavelength of the light rays gets red- or blue-shifted, with red-shift and Cauchy horizon stability when the cosmological horizon has the larger surface gravity. The Cauchy horizon is stable for a range of masses (near to the line OB in figure 6.7) (Mellor and Moss 1990).

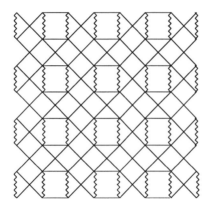

Figure 6.8   A Penrose diagram for a charged black hole in de Sitter space.

In the next chapter we shall relate the surface gravity of the event horizon to the temperature of the black hole. The spacetime cannot be in thermodynamic equilibrium if the surface gravity of the event horizon is different from the surface gravity of the cosmological horizon. Two special cases exist with identical surface gravities. The first is a degenerate case where the surface gravity of both horizons vanishes. This equilibrium is unstable to small perturbations. The second case is when $Q = M$ (the line OD in figure 6.7), when the surface gravity on both horizons becomes $\kappa = \alpha^{-1}(1 - 4GM/\alpha)^{1/2}$, where $\alpha^2 = 3/\Lambda$. This case is thermodynamically stable, in the sense that the horizons have positive specific heats.

The classical stability of this black hole spacetime, albeit with a cosmological constant, means that the basic Einstein theory of general relativity breaks down in a serious way. This leaves us with a challenge: how to formulate quantum theory in the vicinity of a naked singularity.

6.7   The Big Bang

Astronomical observations show that distant galaxies are all receding from us at a rate proportional to their distance away. The constant of proportionality defines the Hubble parameter H_0. Other important cosmological parameters concerning the distribution of matter and the cosmic microwave background are given in table 6.7.

Table 6.4   Observational values of some cosmological parameters. The present-day critical density $\rho_c = 1.9 \times 10^{-26} h^2$ kg m^{-3}, where $h = H_0/(100$ km s^{-1} Mpc).

Quantity		Value
Hubble constant[a]	H_0	87 ± 7 km s^{-1} Mpc^{-1}
Hubble constant[b]	H_0	69 ± 8 km s^{-1} Mpc^{-1}
density parameter (visible)[c]	Ω_{vis}	$0.003\,h^{-2}$
density parameter (kinetic)[d]	Ω	$0.2\,h^{-2}$
cmb temperature[e]	T_0	2.726 ± 0.005 K
cmb fluctuations ($10°$ scale)[f]	δ_T	1.1×10^{-5}

[a] Pierce et al. (1994).

[b] Tanvir et al. (1995).

[c] Faber and Gallagher (1979).

[d] Trimble (1979).

[e] Mather et al. (1993).

[f] Smoot et al. (1992).

Theoretical cosmology is based upon the assumption that when averaged over large enough length scales the universe is isotropic and homogeneous. The isotropy of the universe about our local point of observation is very well established by the isotropy of the cosmic microwave radiation. Homogeneity is very difficult to observe directly because we can only see galaxies from the limited region of space-time joined to us by light rays. It is therefore necessary to take the homogeneity of the universe as an assumption.

According to the singularity theorems, universal expansion implies the existence of the Big Bang singularity, which might be regarded as the origin of spacetime and of matter in the universe.

If we place the homogeneity assumption into the Tolman–Bondi metric by writing $R(r,t) = a(t)r$ and $m(r) = \frac{1}{2}Kr^3$, then the metric takes the Robertson–Walker form,

$$ds^2 = -dt^2 + a^2 \left((1 - Kr^2)^{-1}dr^2 + r^2(d\theta^2 + \sin^2\theta\, d\phi^2)\right). \tag{6.26}$$

By rescaling the radial coordinate it is possible to fix the value of K to be ± 1 or 0. When $K = 0$ the spatial part of the metric is flat. In the $K = 1$ case we can eliminate the coordinate singularity in the metric at $r = 1$ by introducing a new angular variable, $\sin\chi = r$. The spatial part of the metric is identical to the metric on a three-dimensional sphere or hypersphere which is described in appendix D.

The scale factor $a(t)$ satisfies the Friedmann equation (see equation (6.6)),

$$3a^{-2}(\dot{a}^2 + K) = 8\pi G\rho. \tag{6.27}$$

This equation is also valid when the matter content exerts pressure forces. In general, extra equations are also required. The law of conservation of energy or the first law of thermodynamics can be applied. For a comoving region of size a, volume a^3 and total energy $a^3\rho$, we have

$$\frac{d}{dt}(\rho a^3) = -p\frac{d}{dt}(a^3). \tag{6.28}$$

If there is no thermal dissipation, then these two equations together with the equation of state for the matter fields are enough to determine the time evolution of the scale factor. The two most important examples are a radiation-filled universe and a pressure-free universe:

$$p = \tfrac{1}{3}\rho \qquad \rho \propto a^{-3} \tag{6.29}$$

$$p = 0 \qquad \rho \propto a^{-4}. \tag{6.30}$$

Another useful equation that can be derived from the Friedmann and energy conservation equations is

$$\ddot{a} = -\tfrac{8}{3}\pi G\, a(\rho + 3p). \tag{6.31}$$

This equation is independent of K.

The $K = 0$ metric gives the Einstein–de Sitter cosmological model, a flat universe which expands from a curvature singularity and continues expanding for ever, assuming that the density never becomes negative. The $K = -1$ metric also expands for ever from an initial singularity, but if $K = 1$ the universe has a final as well as an initial singularity.

Departures from the Einstein–de Sitter model can be described by the density parameter Ω, defined as the ratio of the density to the critical density ρ_c,

$$\rho_c = 3H^2/8\pi G. \tag{6.32}$$

From equation (6.27) we see that if $K = 0$ then $\Omega = 1$, otherwise $|\Omega - 1|$ increases in proportion to $\dot{a}^{-2}$.

6.8 Anisotropic cosmologies

The Tolman–Bondi metrics provide examples of model universes that are inhomogeneous but isotropic. The contrasting case of models that are homogeneous and anisotropic are known as Bianchi models.

We begin with the hypersphere and choose the special basis of 1-forms ω^i described in appendix C. Then take a metric

$$ds^2 = -dt^2 + \tfrac{1}{4}g_{ij}(t)\omega^i\omega^j. \tag{6.33}$$

The factor of $\tfrac{1}{4}$ is chosen so that the metric reduces to the isotropic case when $g_{ij} = a^2\delta_{ij}$.

The expansion of the spatial sections is described by an expansion rate k and a shear rate s_{ij}. Using 3×3 matrix notation,

$$k = \tfrac{1}{2}\mathrm{tr}(\mathbf{g}^{-1}\dot{\mathbf{g}}) \qquad \mathbf{s} = \tfrac{1}{2}\mathbf{g}^{-1}\dot{\mathbf{g}} - \tfrac{1}{6}\mathrm{tr}(\mathbf{g}^{-1}\dot{\mathbf{g}}). \tag{6.34}$$

The components of the Einstein tensor are listed in table 6.8.

Table 6.5 Components of the Einstein tensor for anisotropic cosmologies.

G	Expression
G_{tt}	$\tfrac{1}{3}k^2 - \tfrac{1}{2}\mathrm{tr}\,\mathbf{s}^2 + 2\mathrm{tr}(\mathbf{g}^{-1}\mathbf{r})$
G_{ij}	$\tfrac{1}{4}\mathbf{g}\left(\dot{\mathbf{s}} - 2\dot{k} + k\mathbf{s} - \tfrac{1}{3}k^2 - \tfrac{1}{2}\mathrm{tr}\,\mathbf{s}^2\right) + \mathbf{r} - \tfrac{1}{2}\mathbf{g}\,\mathrm{tr}(\mathbf{g}^{-1}\mathbf{r})$

The matrix r_{ij} is the Ricci curvature tensor of the spatial sections. This is determined by the properties of the basis ω^i. For our example,

$$\mathbf{r} = K(1 - \mathbf{g}^{-1}\,\mathrm{tr}\,\mathbf{g})(1 - \mathbf{g}\,\mathrm{tr}\,\mathbf{g}^{-1}) \tag{6.35}$$

with $K = 1$. The spatial sections are compact, and the Ricci tensor is positive definite. This case is known as Bianchi type IX. If the same basis forms are used but with χ replaced by $i\chi$, then the Ricci tensor is given by the same result but with $K = -1$. This is Bianchi type VII. Cartesian coordinate forms would give $K = 0$, and this is Bianchi type I. (The other Bianchi types are cases where the metric g has equal eigenvalues.)

The field equations are made up of a set of ordinary differential equations into which we can place any desirable stress–energy tensor. For a perfect fluid with density ρ and pressure p at rest with respect to the expansion,

$$T_{tt} = \rho \qquad T_{ij} = p\, g_{ij}. \tag{6.36}$$

The first of the field equations gives,

$$\tfrac{1}{3}k^2 - \tfrac{1}{2}\mathrm{tr}\ s^2 + 2\mathrm{tr}(\mathbf{g}^{-1}\mathbf{r}) = 8\pi\rho. \tag{6.37}$$

The trace of the spatial equations combined with the first equation gives

$$\dot{k} + \tfrac{2}{3}k^2 + 2\mathrm{tr}\ s^2 = -8\pi(\rho + 3p). \tag{6.38}$$

The behaviour of the solutions to these equations depends critically on the material properties of the fluid. If $\rho > 0$, then the weak energy condition applies. If in addition $\rho + 3p > 0$, then the strong energy condition holds. In the latter case,

$$\dot{k} < -\tfrac{2}{3}k^2. \tag{6.39}$$

Consequently, if the universe is expanding, then there is a time t_0 in the past when $k \to \infty$. The volume of the universe vanishes at $t = t_0$, which may be identified as the singular origin of the universe.

In the special case of isotropy, $k = 3\dot{a}/a$ and equation (6.37) reduces to Friedmann's equation, whilst equation (6.38) becomes equation (6.31). The two equations together are equivalent to Friedmann's equation plus the energy conservation law.

7

Black hole evaporation

It is difficult to appreciate nowadays how unexpected was the first prediction of black hole evaporation. According to classical relativity, black holes are the perfectly stable outcome of gravitational collapse. Prior to Hawking's results (Hawking 1974) it seemed likely that quantised fields would also settle down to a stationary vacuum state and nothing further would happen.

In spacetimes which have a time-like Killing vector it is possible to introduce energy and energy conservation. The vacuum is the state of minimum energy and should therefore be stable. However, in some static spacetimes such as the black hole, the time-like Killing vector is not globally defined. This leads to some subtlety in the choice of quantum states for black hole spacetimes.

7.1 Particle states

The operator approach to quantum field theory provides the simplest way of understanding black hole evaporation. Particle states have to be set up with some care, the first step being to introduce a basis of complex solutions of the wave equation. These solutions have to be divided in some way into positive- and negative-frequency solutions f_i and $\overline{f}_i$. They should also be orthogonal, which means that we need to define a product and impose the relations

$$(f_i, f_j) = -(\overline{f}_i, \overline{f}_j) = \delta_{ij} \quad (f_i, \overline{f}_j) = 0. \tag{7.1}$$

Some of the classical field equations that might be of interest are tabulated in table 7.1. The wave operators satisfy a common identity,

$$\phi_1^* \Delta \phi_2 - \phi_2^* \Delta \phi_1 = -i \mathbf{\nabla} \cdot \mathbf{J} \tag{7.2}$$

for a given vector function $\mathbf{J}(\phi_1, \phi_2)$. For scalar fields,

$$\mathbf{J} = i(\phi_1^* \mathbf{\nabla} \phi_2 - \phi_2^* \mathbf{\nabla} \phi_1). \tag{7.3}$$

When ϕ_1 and ϕ_2 are two solutions of the wave equation, then $\mathbf{\nabla} \cdot \mathbf{J} = 0$. The function $\mathbf{J}$ defines the appropriate inner product by integration on a Cauchy surface Σ,

$$(\phi_1, \phi_2) = - \int_\Sigma d\mu\, \omega(\mathbf{J}) \tag{7.4}$$

where ω is the surface form. A Cauchy surface is a hypersurface that has no edge and time-like curves intersect it only once. The inner product between two solutions of the wave equation is independent of which surface is chosen, because the difference between any two is related to $\nabla \cdot \mathbf{J}$ by the divergence theorem.

Table 7.1   Wave operators for various fields.

Field	Operator Δ
scalar	$-\nabla^2 + \zeta R + m^2$
charged scalar	$-D^2 + m^2$
spinor	$i\boldsymbol{\gamma} \cdot \nabla + m$
vector	$-\delta_a{}^b \nabla^2 + R_a{}^b$
graviton[a]	$\Delta_{L(ab)}^{(cd)}$

[a] See Appendix B for an explanation of this operator.

A free field ϕ is an operator acting on the space of states. It also satisfies the field equation, therefore possessing an expansion

$$\phi = \sum (a_i f_i + a_i^+ \bar{f}_i). \tag{7.5}$$

The coefficients a and a^+ are both operators.

The quantum theory developed in this chapter will be restricted to spacetimes which have a region where there exists a time-like Killing vector $\mathbf{e}_t$. In this region it is possible to write the positive- and negative-frequency solutions $\propto \exp(\pm i\omega t)$.

The coefficients a_i and a_i^+ are then operators which satisfy commutation relations

$$[a_i, a_j^+] = \delta_{ij}. \tag{7.6}$$

The standard quantum theory arguments show that the operator $a_i^+ a_i$ has integer eigenvalues which we associate with the number of particles in each mode. The raising and lowering operators a^+ and a can be associated with creation and annihilation of particles. The vacuum is the zero-particle state, $a_i|0\rangle = 0$ for all a_i.

Interesting effects can arise from the evolution of this vacuum state through regions of spacetime curvature. One possibility is that positive-frequency solutions may evolve into negative-frequency solutions elsewhere. Another possibility, even when the spacetime is stationary, is that the positive-frequency solutions defined in one region may not be regular in another region.

In either case the vacuum state in one flat region will be an excited state in another flat region. This can be a cause of particle production from spacetime curvature.

7.2   Wave propagation on a black hole background

The wave equations on a black hole background are important for understanding a whole range of classical and quantum processes connected with black hole physics. Separation of the wave equations is very important for what follows. This has been described fully by Chandrasekhar (1983), and some details will be omitted here.

For fields with spins higher than 0 we have the choice of working with either field strengths or field potentials. The relevant field strength for the photon is the Maxwell tensor and for the graviton, the Weyl tensor. These field strengths can be converted into fully symmetric SL(2,C) tensors $\Psi^{AB\ldots C}$ by a linear transformation. (This is the same embedding of SO(3,1) into SL(2,C) that we used for fermions in chapter 2.) There are $2s$ indices for a spin-s tensor, and therefore $2s + 1$ independent components of the symmetric tensor, which we label Ψ_n, $n = -s, \ldots, s$.

The field equations for Ψ_n are separable, with

$$\Psi_n = r^{-s} R_n(j, \omega; r) S_n(j, m; \theta) e^{(im\phi - i\omega t)}. \tag{7.7}$$

The angular functions are spin-n spherical eigenfunctions with eigenvalues $j(j + 1)$ and half-integer j (see appendix D). The radial part of the field equation is known as the Teukolski–Starobinskii equation

$$\left(\mathcal{D}^\dagger_{-n/2} \Delta \mathcal{D}_{n/2} + 2(2n - 1)i\omega r \right) R_n = \lambda_n R_n \tag{7.8}$$

where $\lambda_n = (j + n)(j - n + 1)$. The operator $\mathcal{D}_n$ is a radial derivative,

$$\mathcal{D}_n = \frac{\partial}{\partial r} - \frac{i\omega r^2}{\Delta} + n\frac{\Delta'}{\Delta}. \tag{7.9}$$

(For Kerr, the angular functions become spheroidal harmonics. In the radial equations, $-\omega r^2$ is replaced by $am - \sigma_r^2 \omega$ and some restrictions are placed upon n, but the equations are otherwise identical.)

In earlier chapters quantum field theory was developed in terms of field potentials. These are not necessarily the same as the field strengths, but the equations for the potentials can also be separated. The corresponding radial functions will be denoted by $\mathcal{R}_i$. These can be chosen to satisfy scattering equations of standard form,

$$-\frac{d^2 \mathcal{R}_i}{dr^{*2}} + V(r)\mathcal{R}_i = \omega^2 \mathcal{R}_i. \tag{7.10}$$

The coordinate r^* is the tortoise coordinate introduced in chapter 6,

$$r^* = \int dr \, \frac{r^2}{\Delta}. \tag{7.11}$$

The scattering potentials are tabulated in table 7.2. It is not possible to solve the radial equation analytically, but the scattering equations do have some nice properties. The scattering potentials approach zero asymptotically for non-rotating holes, or constants for rotating holes. Near the horizon $r^* \to -\infty$ they decay exponentially, with exponents determined by the surface gravity.

Table 7.2 Scattering potentials for spin s fields (parameterised by f).

Spin	$V(r)$	$f(r)$
0	$r^{-1}f_{,r*} + \lambda_0 r^{-2}f$	Δ/r^2
$\frac{1}{2}$	$\pm\lambda_{1/2}f_{,r*} + \lambda_{1/2}^2 f^2$	$\Delta^{1/2}/r^2$
1	$\lambda_1 f$	Δ/r^4
2	$\pm 6Mf_{,r*} + 36M^2 f^2 + \lambda_2(\lambda_2+2)f$	$r^{-3}\Delta/(\lambda_2 r + 6M)$

Example: Scalar fields

Consider the spherically symmetric Schwarzschild metric with a massless scalar field $\phi(x)$, where the wave equation is

$$\frac{1}{\sqrt{g}} \frac{\partial}{\partial x^\mu} \sqrt{g} g^{\mu\nu} \frac{\partial \phi}{\partial x^\nu} = 0. \tag{7.12}$$

The separable form of solution is

$$\phi = R_{l\omega}(r) Y_{lm}(\theta,\phi) e^{-i\omega t}. \tag{7.13}$$

Substituting this and the metric into the wave equation gives

$$\frac{1}{r^2} \frac{d}{dr} \Delta \frac{d}{dr} R_{l\omega} + \left(\frac{r^2}{\Delta}\omega^2 - \frac{l(l+1)}{r^2} \right) R_{l\omega} = 0. \tag{7.14}$$

This is the same as equation (7.8). We next set $\mathcal{R}_{l\omega} = rR_{l\omega}$, and simplify the derivative terms by introducing the r^* coordinate,

$$\frac{d}{dr} \Delta \frac{d}{dr} R_{l\omega} = \frac{r^3}{\Delta} \frac{d^2 \mathcal{R}_{l\omega}}{dr^{*2}} - \left(\frac{\Delta}{r^2} \right)' \mathcal{R}_{l\omega}. \tag{7.15}$$

Thus

$$\left(-\frac{d^2}{dr^{*2}} + V(r) \right) \mathcal{R}_{l\omega} = \omega^2 \mathcal{R}_{l\omega}. \tag{7.16}$$

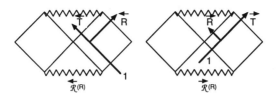

Figure 7.1   Wavevectors of the outgoing and ingoing modes.

The potential $V(r)$ takes the form

$$V(r) = j(j+1)\frac{\Delta}{r^4} + \frac{\Delta}{r^3}\left(\frac{\Delta}{r^2}\right)'. \tag{7.17}$$

This appears as the first line in table 7.2.

7.3   Black hole vacuum states

We are in a position now to begin the first step towards finding a basis of particle states. The wave equations have incoming and outgoing wave solutions with asymptotic form

$$f_i = e^{-i\omega t \pm i\omega r^*}Y_{lm}(\theta, \phi) \tag{7.18}$$

which can be combined to form a basis of solutions. The most important task of this section will be separating the basis of solutions into positive and negative frequency.

The radial mode equations define a scattering problem in one dimension. The solutions can be divided into ingoing and outgoing modes as indicated in figure 7.1, with reflection and transmission coefficients R and T. For example,

$$\overrightarrow{\mathcal{R}}_{\omega l} \rightarrow \begin{cases} e^{i\omega r^*} + \overrightarrow{R}_{\omega l}e^{-i\omega r^*} & r \rightarrow r_h \\ \overrightarrow{T}_{\omega l}e^{i\omega r^*} & r \rightarrow \infty \end{cases} \tag{7.19}$$

emerges from the past horizon with unit amplitude, and

$$\overleftarrow{\mathcal{R}}_{\omega l} \rightarrow \begin{cases} \overleftarrow{T}_{\omega l}e^{i\omega r^*} & r \rightarrow r_h \\ e^{-i\omega r^*} + \overleftarrow{R}_{\omega l}e^{i\omega r^*} & r \rightarrow \infty \end{cases} \tag{7.20}$$

comes in from infinity with unit amplitude.

Since the spacetime is asymptotically flat, the natural vacuum state to use far from the hole is the one constructed from positive-ω solutions. A possible problem with this state is that the time coordinate t diverges on the black hole horizon. In fact, in this state the regularised expectation value of the energy density diverges on the event horizon.

The appropriate physical states to use for a black hole spacetime should lead to regular expectation values on the future horizon $\mathcal{H}^+$. An obvious procedure is to try using positive-frequency Kruskal coordinate modes to define the vacuum state. The modes given below are constructed by combining t modes from regions I and III of the complete spacetime,

$$\overrightarrow{f}_\omega^{(U)} = (e^{\pi\omega/2\kappa}\overrightarrow{f}_\omega^{(R)} + e^{-\pi\omega/2\kappa}\overrightarrow{f}_\omega^{(L)})/[2\sinh(\pi\omega/\kappa)]^{1/2}. \qquad (7.21)$$

An extra superscript distinguishes modes $f_\omega^{(L)}$ which vanish in region I from $f_\omega^{(R)}$ which vanish in region III.

The particular combination is determined by the requirement that the modes on the past horizon have to be analytic and bounded as functions of U in the lower complex U plane. These conditions imply that the Fourier transforms with respect to U vanish when ω is negative.

The only problem preventing regularity is a branch cut where $\arg U = -\pi/2$. From the definition of U,

$$e^{-i\omega(t-r^*)} = \begin{cases} e^{i\omega\log(-U)/\kappa} & \Re U < 0 \\ e^{i\omega\log(U)/\kappa} & \Re U > 0. \end{cases} \qquad (7.22)$$

On the past horizon where $V = 0$ and $r = r_h$,

$$f_\omega^{(L)} = 0 \qquad f_\omega^{(R)} \to \text{const} \times e^{-\pi\omega/2\kappa} \quad \Re U < 0 \qquad (7.23)$$

$$f_\omega^{(R)} = 0 \qquad f_\omega^{(L)} \to \text{const} \times e^{\pi\omega/2\kappa} \quad \Re U > 0 \qquad (7.24)$$

as $\arg U \to -\pi/2$. There is no jump in $\overrightarrow{f}_\omega^{(U)}$, no branch cut and therefore the modes are analytic.

The same procedure can be applied for positive-frequency modes on $\mathcal{H}^-$ with the V coordinate. This leads to a selection of vacuum states depending on the choice of modes for outgoing and ingoing waves (Unruh 1976):

1. Boulware vacuum $|B\rangle$ using $\{\overrightarrow{f}^{(R)}, \overleftarrow{f}^{(R)}, \overrightarrow{f}^{(L)}, \overleftarrow{f}^{(L)}\}$.
2. Unruh vacuum $|U\rangle$ using $\{\overleftarrow{f}^{(R)}, \overleftarrow{f}^{(L)}, \overrightarrow{f}^{(U)}\}$.
3. Hartle–Hawking vacuum $|H\rangle$ using $\{\overleftarrow{f}^{(V)}, \overrightarrow{f}^{(U)}\}$.

The Boulware vacuum defines the reference vacuum state for observers far from the hole. However, the Unruh vacuum state is regular on the future event horizon and the correct ground state for black holes that form from gravitational collapse. The Hartle–Hawking vacuum is regular on both $\mathcal{H}^+$

and $\mathcal{H}^-$ and applies to the time-symmetric problem of a black hole in thermal equilibrium.

A black hole which forms from gravitational collapse has a modified conformal diagram shown in figure 7.2. In the asymptotic regions the wave equation decomposes as before, except that now it is preferable to transform $\mathcal{R}_{l\omega}(r)$ to $\mathcal{R}_l(r, t)$. The diagram shows that a component of the outgoing modes near the event horizon has emerged from the collapsing star after travelling from past infinity. These modes are crucial because they turn out to be positive frequency in the Kruskal coordinate.

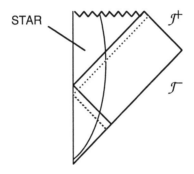

Figure 7.2 Collapsing star.

As the critical waves enter the spacetime they are positive frequency in the asymptotic time coordinate. The high-frequency modes can be followed in the geometrical optics limit, $\mathcal{R} = A\exp(i\Phi(u, v))$, with parallel propagated wavevector $\mathbf{k} = \nabla\Phi$. Ordinary, flat-space, propagation in the Penrose diagram corresponds to parallel propagation on the actual spacetime. We therefore follow the wavevector through from $\mathbf{k} = -\omega e_v$ on past infinity, to $\mathbf{k} = -\omega e_U$ on future infinity, where U is the Kruskal coordinate. The actual phase is therefore $\Phi = cU + d$, where $c > 0$ and d are constants, and the waves are Kruskal U positive frequency, the same as the Unruh vacuum.

7.4   Particle fluxes

The Unruh vacuum state is not the state which registers zero on a particle detector at rest. In this state there is a flow of energy out of the hole. A simple but effective way to demonstrate the existence of the flux is to find the quantum expectation value of the stress–energy tensor.

The stress–energy operator for a massless field has the form

$$T^{ab} = \nabla^{(a}\phi\nabla^{b)}\phi - \tfrac{1}{2}g^{ab}(\nabla^c\phi)(\nabla_c\phi). \tag{7.25}$$

Substituting the field expansion into the vacuum expectation value for T^{ab} gives

$$\langle T^{ab}\rangle = \sum_i T^{ab}(f_i, \overline{f}_i) \tag{7.26}$$

where

$$T^{ab}(u, v) = \nabla^{(a}u\nabla^{b)}v - \tfrac{1}{2}g^{ab}(\nabla^c u)(\nabla_c v). \tag{7.27}$$

In general this expression is infinite and a regularisation procedure has to be defined to obtain a finite result.

From the flux,

$$\langle T^{rt}\rangle = \sum_i T^{rt}(\overleftarrow{f}_i, \overleftarrow{f}_i) + \sum_i T^{rt}(f_i^{(U)}, f_i^{(U)}). \tag{7.28}$$

Using the asymptotic forms of the wave modes we have

$$T^{rt}(\overleftarrow{f}_i, \overleftarrow{f}_i) \rightarrow \frac{1}{2\pi r^2}|Y_{lm}(\theta, \phi)|^2\omega(1 - |R_{l\omega}|^2) \tag{7.29}$$

$$T^{rt}(\overrightarrow{f}_i, \overrightarrow{f}_i) \rightarrow -\frac{1}{2\pi r^2}|Y_{lm}(\theta, \phi)|^2\omega|T_{l\omega}|^2. \tag{7.30}$$

It follows from the form of the modes $f^{(U)}$ that

$$\langle T^{rt}\rangle \rightarrow -\frac{1}{2\pi r^2}\sum_l P_l(\cos\theta)^2 \int_0^\infty d\omega |T_{l\omega}|^2 \frac{\omega}{e^{2\pi\omega/\kappa} - 1}. \tag{7.31}$$

This represents a thermal flow of energy, weighted by the transmission amplitude $T_{l\omega}$. The total flux F_j from any particle species j follows from the Stefan–Boltzmann formula with area $16\pi M^2$ and temperature $\kappa/2\pi$,

$$F_j = 16\pi M^2 a_s \Gamma_j (\kappa/2\pi)^4 = \alpha_j M^{-2} \tag{7.32}$$

where a_s is the Stefan–Boltzmann constant and Γ_j is a weighted transmission coefficient which depends on the particle mass and spin. Numerical estimates for massless particles give $\alpha_1 = 2.8 \times 10^{32}$ J kg^2 s^{-1} for spin 1 and $\alpha_{1/2} = 2.5\alpha_1$ for spin $\tfrac{1}{2}$ (using Page 1978). Total values of α obtained by summing over particles up to a given rest mass in the standard model are tabulated in table 7.3.

Table 7.3 Black hole lifetimes and luminosities.

Temperature	α/α_1	Lifetime (s)	Mass (kg)
1 MeV	27	4.7×10^{21}	1.06×10^{13}
100 MeV	27	4.7×10^{15}	1.06×10^{11}
1 GeV	178	7.2×10^{11}	1.06×10^{10}
100 GeV	218	5.9×10^{5}	1.06×10^{8}
1 TeV	257	5.3×10^{2}	1.06×10^{7}
100 TeV	287	5.1×10^{-4}	1.06×10^{5}

Energy conservation requires that the energy flux be balanced by loss of mass from the hole,

$$\frac{dM}{dt} = -\alpha M^{-2}. \tag{7.33}$$

Consequently, if α is taken to be constant, the hole has a lifetime $M^3/3\alpha$. The hole becomes hotter as it evaporates and behaves as if it had a negative specific heat.

Residual lifetimes for black holes of various masses are tabulated in table 2. The radiation from these holes would show up in the γ-ray background. This has been used to set upper limits of around 10^4 black holes per cubic parsec in this mass range. The γ-rays would have a frequency spectrum proportional to ν^{-2} (Page and Hawking 1976, Carr 1976).

At energies above 200 MeV the black hole can radiate free quarks. Calculations suggest that the hole becomes surrounded by a fireball containing a free quark phase of matter, as yet unobserved on Earth (Moss 1985). The γ-ray emission of this fireball is greatly enhanced and could lead to a pronounced bump in the γ-ray spectrum. Above 200 MeV, the free quarks radiated by the black hole lead to showers of pions and other hadrons similar to the jet events observed in particle accelerators. At still higher energies the black hole encounters physics beyond what can be achieved by particle accelerators on the Earth.

A black hole can carry charges associated with long-range forces, in particular, with electric and magnetic fields. The effects of large electric fields on charged particles can lead to pair creation. Charged black holes therefore radiate a combination of a thermal and a non-thermal charged flux.

For $Q = M$, the temperature vanishes and there is no thermal radiation. However, pair creation of charged particles can discharge the hole even in this case and lead to eventual evaporation. The charged particle production is influenced by the presence of the vector potential, $\mathbf{A} = Q(r^{-1} - r_+^{-1})\mathbf{dt}$, chosen here in a form that is regular on the event horizon. This acts as an effective chemical potential $\mu = A_t/e$, for particles of charge e (see chapter

5). The charged modes with frequency ω have wavenumber $k = \omega + \mu$. Modes with frequency $-\mu < \omega < 0$ correspond to real particles at infinity (if $eQ < 0$) and they are produced even when the temperature vanishes.

There may be cases where the particle theory contains no light charged particles with the same charge that the hole possesses: for example, if the hole carries a magnetic charge and its mass is smaller than a free magnetic monopole. In these cases an extremal black hole can live forever.

7.5   Thermodynamics

One consequence of black hole radiation that really catches the imagination is the whole area of the thermodynamic properties of black holes. This includes the existence of an entropy associated with the black hole, a wholly revolutionary concept.

The first estimate of the black hole entropy was due to Beckenstein (1974). The authors of the laws of black hole mechanics (Bardeen et al. 1973) considered the thermodynamic analogy, but they argued that it was only an analogy. Reviews of more recent work can be found in Brown and York (1993a,b) and Wald (1994).

The laws of black hole thermodynamics begin with the fact that the outward-going flux calculated in the previous section is exactly equal to the absorption rate from a gas in thermal equilibrium at the black hole temperature. This is expressed by the zeroth law:

(0) A black hole can be in thermal equilibrium with radiation at the Hawking temperature.

The other laws follow from classical black hole mechanics:

(1) Under external influences,

$$\delta M = \tfrac{1}{4}(\kappa/2\pi G)\delta\mathcal{A} + \text{ work terms.} \qquad (7.34)$$

(2) The area increase law: $\delta\mathcal{A} \geq 0$.

Recognising that M is the total energy and that the Hawking temperature $T = \kappa/2\pi$, we associate the entropy with the area of the event horizon

$$\mathcal{S} = \tfrac{1}{4}\mathcal{A}/(G\hbar). \qquad (7.35)$$

If this entropy is neglected, then violations of the second law of thermodynamics can be manufactured.

For detailed thermodynamical calculations we would like to calculate the partition function. The partition function for quantum fields in the grand canonical ensemble can generally be related to a path integral by analytic continuation in the time coordinate. The particle propagators on the new spacetime, now Riemannian, define thermal Green functions on the original spacetime.

Analytical continuation can also be used to obtain the thermal Green function on a black hole spacetime (Hawking and Israel 1979). In this case we set

$$U = -Re^{i\Theta} \tag{7.36}$$

$$V = Re^{-i\Theta} \tag{7.37}$$

then

$$ds^2 = \kappa_h^{-2}(1 - 2GM/r)R^{-2}(dR^2 + R^2 d\Theta^2) + r^2(d\theta^2 + \sin^2\theta\, d\phi^2). \tag{7.38}$$

This metric is regular at $r = 2GM$ and covers the region $2GM \leq r < \infty$ provided that $\Theta = i\kappa t$ has period 2π. The Euclidean time coordinate $\tau = it$ has period $2\pi/\kappa$.

Green functions on this space continue to thermal Green functions in the original space, equivalent to Green functions in the Hartle–Hawking vacuum state. The reason that it corresponds to this state and not one of the others is because the state is regular on the two-sphere $R = 0$ which continues to both horizons $\mathcal{H}^+$ and $\mathcal{H}^-$.

This result is very powerful because periodicity in imaginary time completely characterises thermal Green functions. We can immediately deduce that, far from the hole, the matter fields are thermal even in the presence of interactions.

We can now calculate a partition function from a path integral on a region of the Riemannian black hole background,

$$Z(\beta, \mu) = \int d\mu[\mathbf{g}, \phi]\, e^{-I/\hbar}. \tag{7.39}$$

This should be able to represent a black hole and its radiation fields in the grand canonical ensemble confined to some finite volume.

At first, it becomes difficult in this way to find any term corresponding to the black hole entropy. *It is necessary to include the gravitational action in the definition of the partition function.* In other words, the black hole entropy is an effect of quantum gravity and the partition function includes a sum over metrics that are periodic with period β.

To leading order in $\hbar$, the partition function is simply related to the action of the black hole solution, $Z = \exp(-I/\hbar)$. The next order includes the contribution from the heat bath around the black hole. The following method of calculating the action, based upon a canonical decomposition, generalises easily to all types of black hole solution.

The gravitational action for a region $\mathcal{M}$ with boundary $\mathcal{B}$ is

$$I = \frac{1}{16\pi G} \int_{\mathcal{M}} d\mu\, R + \frac{1}{8\pi G} \int_{\mathcal{B}} d\mu\, (k - k_0). \tag{7.40}$$

The first boundary term is explained in appendix B. The second boundary term, which has no effect on the field equations, can be used to make the action finite whatever the size of the box.

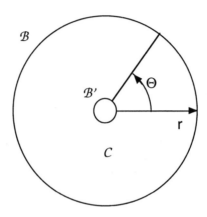

Figure 7.3   Riemannian (r, t) section of a black hole with boundary $\mathcal{B}$

The symmetry of the metric under time translation can be exploited by splitting the volume integral up into integrals over surfaces of constant time. The topology of the Riemannian manifold is not suited to this decomposition, but this can be overcome by introducing an annulus $\mathcal{C}$, shown in figure 7.3, with action I_c,

$$I_c = \frac{1}{16\pi G} \int_{\mathcal{C}} d\mu\, R + \frac{1}{8\pi G} \int_{\partial\mathcal{C}} d\mu\, k. \tag{7.41}$$

In the limit that $\mathcal{B}' \to \mathcal{H}$,

$$I = I_c + \frac{1}{8\pi G} \int_{\mathcal{B}} d\mu\, k_0 - \frac{1}{8\pi G} \int_{\mathcal{H}} d\mu\, k. \tag{7.42}$$

Canonical decomposition of the action I_c is described in appendix C and gives

$$I_c = \frac{1}{16\pi G} \int_{\Sigma} d\mu\, dt\, (\dot{g}_{ij} p^{ij} - N\mathcal{H} - N^i \mathcal{H}_i) + \frac{1}{8\pi G} \int_{\partial\Sigma} d\mu\, dt\, \widehat{k}. \tag{7.43}$$

The factor $\widehat{k}$ is the extrinsic curvature of $\partial\Sigma$ embedded in Σ. There are two components of $\partial\Sigma$, but the contribution from the component in $\mathcal{B}'$ vanishes as $\mathcal{B}' \to \mathcal{H}$.

Time translation symmetry implies that $\dot{g}_{ij} = 0$ and the constraint equations are $\mathcal{H} = \mathcal{H}_i = 0$, therefore only the surface term in I_c survives. Returning to the action I we now have

$$I = \frac{\hbar\beta}{8\pi G}\int_{\partial\Sigma\cap\mathcal{B}} d\mu\,(\widehat{k} - Nk_0) - \frac{1}{8\pi G}\int_{\mathcal{H}} d\mu\,k. \qquad (7.44)$$

The extrinsic curvatures are given by the divergence of the respective normal vectors. For Schwarzschild in the infinite volume limit,

$$I = \hbar\beta M - \tfrac{1}{4}A/G. \qquad (7.45)$$

The calculation generalises in this form very easily to charged or rotating black holes. The only change is that the total charge enters the boundary term in equation (7.43) and the angular momentum enters into the divergence of the normal vectors when evaluating $\widehat{k}$ on $\mathcal{B}$.

The thermodynamic identity for the entropy is $\mathcal{S} = \partial\Omega/\partial T$, where Ω is the thermodynamic potential $\Omega = -T\log Z$. This agrees with the result $\mathcal{S} = A/(4G\hbar)$.

7.6   The final state of black hole evaporation

The black hole evaporation calculation fails for masses as low as 10 μg because the radiation density is comparable to the density of the hole and the quasistatic assumptions fail. The ultimate fate of a black hole is therefore an open question.

If a hole evaporates away completely then one possibility for the global spacetime geometry would resemble figure 7.4. An immediate problem is raised by the absence of a global Cauchy surface. The product of two solutions to the wave equation depends on whether it is evaluated before or after the hole has evaporated (on surfaces $\mathcal{C}_1$ or $\mathcal{C}_2$ in figure 7.4).

An alternative possibility is that the singularity never forms and instead there is the formation of a disconnected baby universe, as shown in figure 7.5. This kind of topology change can happen only in non-Lorentzian spacetimes, but we have already seen how Riemannian metrics can be used to find the partition function for quantum fields, and how the gravitational action is related to the black hole entropy. It may be that Riemannian metrics contribute substantially to quantum effects on Planck scales. The tubes connecting different universes have come to be known as wormholes.

Wormholes do not exist as solutions of the Einstein equations for pure gravity even with Riemannian signature. With matter sources they arise easily.

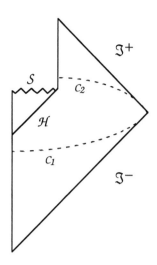

Figure 7.4   Penrose diagram of an evaporating hole. Two partial Cauchy
surfaces $\mathcal{C}_1$ and $\mathcal{C}_2$ are shown.

A simple example consists of an SU(2) Yang–Mills field **A** coupled to gravity.
The metric ansatz for a space with the topology R$\times$S^3

$$ds^2 = dt^2 + a(t)^2 d\Omega^2 \tag{7.46}$$

can be used, with

$$A = \mathbf{T}_i \boldsymbol{\omega}^i \quad d\boldsymbol{\omega}_3^2 = \tfrac{1}{4} \sum \bar{\boldsymbol{\omega}}^i \otimes \boldsymbol{\omega}^i. \tag{7.47}$$

The $\mathbf{T}_i$ are a Hermitian SU(2) Lie algebra basis and $\boldsymbol{\omega}^i$ are the invariant forms
on S^3 defined in appendix C.

The Einstein–Yang–Mills equations have a solution $a = \sqrt{r_0^2 + t^2}$, where
$r_0 = 4\pi G/e^2$. The metric approaches two flat-space regions for $t \rightarrow \infty$ and
$t \rightarrow -\infty$. These regions are joined by a throat which has a minimum radius
r_0.

It is possible to construct solutions in which many wormhole exits exist in
the same universe. Take semi-wormholes (i.e. $t < 0$) opening on to the same
space and identify pairs of throats at $t = 0$. This may be a picture for the
structure of spacetime on very small scales. An interesting possibility that
this raises is that the gauge fields on the throats need only be identical up
to gauge transformations. As a consequence the gauge symmetry is broken
without the need for a Higgs field.

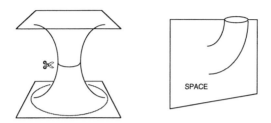

Figure 7.5   A wormhole connecting flat regions and how it may be cut to reveal a baby universe.

7.7   de Sitter space

Although de Sitter space is primarily a cosmological model it also has a set of coordinates where the metric has many features of a black hole (see section 6.5). The event horizon has surface gravity $\kappa_h = \alpha^{-1}$. Fields on this space must therefore have a time-symmetric quantisation, with mode functions constructed as before, and a temperature $\hbar\kappa_h/2\pi$ (Gibbons and Hawking 1977).

The Riemannian section of de Sitter space is a four-sphere of radius α. Green functions on the four-sphere can be continued to de Sitter space and correspond to thermal Green functions, with de Sitter invariance properties and time reversal symmetry. Because of the symmetry of de Sitter space it is simple to find explicit formulae for these Green functions.

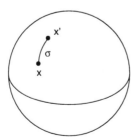

Figure 7.6   The geodesic distance σ on the four-sphere section of de Sitter space.

The symmetry implies that it is possible to write the scalar Green function $G_E(\mathbf{x}, \mathbf{x}')$ in terms of the distance $\sigma(\mathbf{x}, \mathbf{x}')$ along the shortest geodesic from $\mathbf{x}$ to $\mathbf{x}'$ (figure 7.6). This distance has the property that $\sigma_{;\mu}\sigma^{;\mu} = 1$. On the four-sphere it is easily seen (by explicit calculation or otherwise) that it also satisfies

$$\sigma_{;\mu\nu} = \alpha^{-1}\cot(\sigma/\alpha)(g_{\mu\nu} - \sigma_{;\mu}\sigma_{;\nu}) \tag{7.48}$$

$$\sigma_{;\mu'\nu} = -\alpha^{-1}\mathrm{cosec}(\sigma/\alpha)(g_{\nu\mu'} - \sigma_{;\mu'}\sigma_{;\nu}) \tag{7.49}$$

where $g_{\nu\mu'}$ is known as the bitensor of parallel transport, and satisfies $g_{\nu\mu'}\sigma^{;\mu'} = -\sigma_{;\nu}$. The first equation allows the propagator equation to be written

$$G'' + 3\alpha^{-1}\cot(\sigma/\alpha)G' - m^2 G = 0 \tag{7.50}$$

when $\sigma > 0$, with $G(\sigma) \to 1/(2\pi^2\sigma^2)$ as $\sigma \to 0$. The solution is a hypergeometric function

$$G(\sigma) = \frac{1}{16\pi\alpha^2}(\tfrac{1}{4} - \nu^2)\sec(\pi\nu)F(\tfrac{3}{2} + \nu, \tfrac{3}{2} - \nu, 2; \cos^2(\sigma/2\alpha)) \tag{7.51}$$

where $\nu^2 = \tfrac{9}{4} - \alpha^2 m^2$.

The de Sitter version of the propagator can be found by replacing $\sigma(\mathbf{x}, \mathbf{x}')$ with the distance $l(\mathbf{x}, \mathbf{x}')$ in the embedding Minkowski spacetime. (The embeddings are given in section 6.5.) Then

$$G_H(\mathbf{x}, \mathbf{x}') = \frac{1}{16\pi\alpha^2}(\tfrac{1}{4} - \nu^2)\sec(\pi\nu)F(\tfrac{3}{2} + \nu, \tfrac{3}{2} - \nu, 2; 1 - \tfrac{1}{2}l^2/\alpha^2). \tag{7.52}$$

The embedding coordinates are related to the Robertson–Walker time coordinate through $\cosh(t/\alpha)$; therefore it is evident that the propagator is periodic in imaginary time with period $2\pi\alpha$. It cannot be used as a Feynman propagator, but represents a canonical ensemble with a temperature $\hbar/(2\pi\alpha)$. The propagator is used for calculating the amplitude of density fluctuations in inflationary cosmological models in chapter 8.

7.8 Gravitational instantons

The suggestion that Riemannian metrics might contribute to quantum fluctuations on very small scales was mentioned earlier (section 7.6). Solutions to the Einstein equations with finite action might be taken to represent quantum tunnelling and are therefore particularly interesting. These are called gravitational instantons, a term properly reserved for vacuum solutions of the Einstein equations, although more general solutions can also be considered.

The four-sphere is the simplest example of a gravitational instanton. Its use in describing the quantum origin of the universe will be discussed in chapter

9. This instanton has also been used to try and explain why the present value of the cosmological constant is zero. The path integral contribution from this instanton increases with decreasing values of the cosmological constant. Therefore low values of the cosmological constant will dominate in any theory where quantum amplitudes include a sum over fields which somehow induce different values of the cosmological constant.

Table 7.4 Some gravitational instantons.

Space	Source	χ	τ	Action
S^4	Λ	2	0	$-\pi\alpha^2$
CP^2	Λ	3	1	$-3\pi\alpha^2/4$
$S^2 \times S^2$	Λ	4	0	$-2\pi\alpha^2/3$
$CP^2 \# CP^2$	Λ	4	0	$-0.95 \times 2\pi\alpha^2/3$
K3	0	24	-16	0
$S^2 \times S^2$	Λ, F	4	0	$-\pi\alpha^2 + 2\pi\alpha M$

Some other compact instantons are shown in table 7.4. The Euler number χ and the Hirtzbruch signature τ,

$$\chi = \frac{1}{128\pi^2} \int_{\mathcal{M}} \epsilon_{ab}{}^{ef} \epsilon_{cd}{}^{gh} R^{abcd} R_{efgh} \tag{7.53}$$

$$\tau = -\frac{1}{96\pi^2} \int_{\mathcal{M}} \epsilon_{ab}{}^{ef} R^{abcd} R_{efcd} \tag{7.54}$$

help characterise the spaces, though not uniquely.

The space CP^2 has the Fubini study metric

$$ds^2 = (1 - \tfrac{1}{2}r^2/\alpha^2)^{-2}(dr^2 + r^2 \omega^3 \otimes \omega^3) + (1 - \tfrac{1}{2}r^2/\alpha^2)^{-1}r^2(\omega^1 \otimes \omega^1 + \omega^2 \otimes \omega^2) \tag{7.55}$$

where ω^i are the invariant forms on S^3. This space has the peculiarity that it is not possible to define spinors consistently on the complete space, and its physical usefulness may be severely restricted for this reason.

The space $S^2 \times S^2$ in the case of gravity and cosmological constant is the product of two geometrical spheres. It is also the analytic continuation of the metric of a black hole in de Sitter space in the limit $M^2 = \alpha^2/27$ (the point A in figure 6.7). In this limit the two spacetime horizons merge and both the surface gravities vanish.

The corresponding limit of a rotating hole in de Sitter space when analytically continued gives the Page metric (Page 1978) on an S^2 fibre bundle over S^2. No explicit form exists so far for the Einstein metric on K3.

The charged black hole with magnetic charge $P = M$ also gives an instanton which is listed in the table. The two horizons have equal, though in this case non-zero, surface gravities. This condition is necessary because the periodicity of the time coordinate on the Riemannian space is constant and related to the surface gravity of either horizon. The electrically charged and the rotating analogues would also produce regular instantons except that some of the field components become complex. Arguments for using both real and complex instantons in quantum gravity will be presented in chapter 9.

8

The inflationary universe

Inflationary models offer the best hope yet of explaining most of the large-scale features of the universe. These features reflect, as in all cosmological models, a combination of the initial conditions at the Big Bang and the subsequent evolution of the universe. In inflationary models however, most of the large-scale structure is determined by the developing universe and therefore it is, at least in principle, predictable.

Inflation precedes a conventional hot Big Bang plasma phase. Some of the major stages of the hot Big Bang are listed in figure 8.1. Where we place inflation depends upon various factors. An important consideration is at what stage the baryon asymmetry of the universe was generated. Electroweak or grand unification energy scales are the favourites. Inflation reduces this asymmetry to zero, to all intents and purposes, and must take place before baryogenesis.

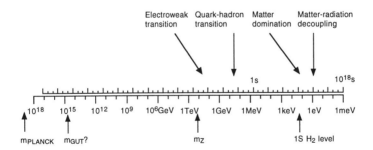

Figure 8.1 Stageposts in the hot Big Bang universe. The upper scale is time and the lower, energy.

In the original inflationary models of Guth (Guth 1981) and some of the early models (Linde 1982, Albrecht and Steinhart 1982, Hawking and Moss 1982), inflation was a result of a supercooled phase transition associated with a Higgs field. In more recent times model-builders have found it necessary to introduce one or more scalar fields whose sole purpose is to drive the inflationary phase.

8.1 The inflationary universe

A period of inflation has many important consequences for the subsequent evolution of the universe. In the first place inflation is capable of producing a homogeneous universe from a very wide range of initial conditions. This is described by a series of cosmological no-hair theorems because the process involved is analogous to the way that gravitational collapse approaches the black hole state.

Also of considerable importance is the fact that physical processes act over distance scales that would normally be causally disconnected in the hot Big Bang. This is the effect that allows predictions of the large-scale structure in inflationary models.

The maximum range of physical phenomena in the early universe defines a physical horizon scale λ_C. The size of the physical horizon in a Friedmann model is set by the velocity of light and the local expansion timescale $1/H$, where H is the expansion rate. Features in the large-scale structure of size λ_L scale with the scale factor a. The expansion rate H is given by $\dot{a}/a$, therefore the ratio λ_L/λ_C is equal to $\dot{a}$ and increases only if $\ddot{a} > 0$. We will use this condition as the definition of inflation, at least as far as homogeneous models are concerned. The Friedmann equation

$$\ddot{a} = -\tfrac{8}{3}\pi G(\rho + 3p)a \tag{8.1}$$

implies that inflation can only take place when $\rho + 3p < 0$.

Possibly the most important aspect of inflation is that it enables us to explain the origin of the density perturbations which lead to galaxies and galaxy clusters. During inflation, a comoving length scale that corresponds to a galaxy today is smaller than the physical horizon and quantum fluctuations can cause inhomogeneities. At a critical time during the inflationary era the embryonic galaxy leaves the physical horizon only to re-enter at a much later epoch during the hot Big Bang. If the scale $\lambda_L = 2\pi a/k$ for fixed k, then it crosses the horizon and $H\lambda_L = 1$ at the times t_k when

$$H(t)a(t) = k/2\pi. \tag{8.2}$$

This is displayed in figure 8.2.

An important consequence of inflation for observational tests is the smallness of the spatial curvature. The density parameter Ω, the ratio of the density of the universe to the density of a flat universe with the same

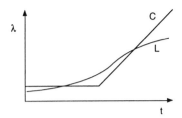

Figure 8.2   A comoving galaxy scale (L) and the causal horizon (C).

expansion rate, satisfies

$$|\Omega - 1| = (Ha)^{-2} \tag{8.3}$$

from the Friedmann equation and approaches $\Omega = 1$ during inflation. A very long time after the end of inflation the value of the density parameter will begin to drift away from unity. If this happens at the present epoch then the present epoch is a special one, which runs contrary to the spirit of the Copernican principle. In this sense it is often said that the inflationary models predict $\Omega = 1$ today.

8.2   Inflationary models of the early universe

The time development of the universe during inflation can be seen most simply in homogeneous and isotropic cosmological models with matter content dominated by a scalar field $\phi(t)$. The density and pressure of the scalar field with potential $V(\phi)$ are

$$\rho = \tfrac{1}{2}\dot{\phi}^2 + V(\phi) \quad p = \tfrac{1}{2}\dot{\phi}^2 - V(\phi). \tag{8.4}$$

The Friedmann equation for the scale factor is then

$$3a^{-2}(\dot{a}^2 + 1) = 8\pi G\rho. \tag{8.5}$$

The energy conservation equation is equivalent to the scalar field equation

$$\ddot{\phi} + 3H\dot{\phi} + V'(\phi) = 0 \tag{8.6}$$

where prime denotes derivative with respect to ϕ.

The condition for inflation, $\rho + 3p < 0$, implies that $\dot{\phi}^2 < V(\phi)$. This can be achieved when the scalar field equation is overdamped, i.e. when the $\ddot{\phi}$ term can be neglected and $\dot{\phi} \sim V'/H$. This is often called a slow rolldown. Since $H^2 > GV$, the condition for slow rolldown inflation is

$$G^{-1/2}(V'/V) \ll 1. \tag{8.7}$$

The simplified equations that describe the slow rolldown phase are

$$\dot{\phi} = -\frac{V'}{3H} \qquad H^2 = \frac{8\pi G}{3}V \qquad \dot{a} = Ha. \tag{8.8}$$

Inflation will end when these necessary conditions break down. Then energy transferred from the scalar field to radiation can come to dominate the energy density of the universe. This process is known as reheating, and represents the source of all energy and matter in the present-day universe.

Inflationary models are constrained in practice by the duration of the inflationary era, which should be large enough for the universe to have the possibility of reaching its present size. They are also constrained to have adequate reheating for the hot Big Bang model and to have density fluctuations of the right size.

Example: Chaotic inflation

Even the simple massive scalar field with a quadratic potential $V(\phi) = \frac{1}{2}m^2\phi^2$ can bring about inflation. The slow rolldown equations (8.8) are consistent with the condition $\dot{\phi}^2 < V(\phi)$ when $\phi > \phi_c = (12\pi G)^{-1/2}$. The slow rolldown solution for $\phi(t)$ is simply

$$\phi(t) = \phi_c(1 - m(t - t_c)) \tag{8.9}$$

where $t = t_c$ marks the end of inflation, and the scale factor is given by

$$a(t) = a_c \exp\left[\frac{1}{3}\left(1 - \frac{\phi^2}{\phi_c^2}\right)\right]. \tag{8.10}$$

8.3   The origin of structure

An exciting aspect of the inflationary scenario is that it enables us to explain the origin of structure in the universe from quantum fluctuations during inflation. Calculations of the size and spectrum of these fluctuations is an exacting test for the formulation of quantum field theory in curved spacetime.

The scalar field will now take the form $\phi(t) + \delta\phi(\boldsymbol{x}, t)$, where $\delta\phi$ denotes the fluctuating part. Insofar as the Hubble parameter H varies slowly during inflation, we will be able to regard $\delta\phi$ as a free quantum field on de Sitter space. The propagator of a field on de Sitter space is calculated in chapter 7.

This propagator has important limits: the regularised coincident limit,

$$G(\mathbf{x}, \mathbf{x})|_{\mathrm{reg}} = \frac{3H^4}{8\pi^2 m^2} \tag{8.11}$$

and also

$$G(\mathbf{x}, \mathbf{x}') \sim \begin{cases} \dfrac{3H^4}{8\pi^2 m^2} - \dfrac{H^2}{4\pi^2}\log(Hl) & \text{for } Hl \gg 1 \\[2mm] \dfrac{3H^4}{8\pi^2 m^2} - \dfrac{H^2}{16\pi} & \text{for } Hl \approx 1 \end{cases} \tag{8.12}$$

where $m \ll H$ and l is the proper distance between $\mathbf{x}$ and $\mathbf{x}'$.

Fluctuations with scales $Hl \gg 1$ are outside of the physical horizon. For density fluctuations on these scales there is a gauge ambiguity that needs to be resolved. The quantity $\delta = \delta\rho/(p + \rho)$ turns out to be a reliable measure of the amplitude, as will be demonstrated in the next section. For the scalar field, $\delta\rho = V'\delta\phi$ and $p + \rho = \dot{\phi}^2$. Therefore, in the slow rolldown limit,

$$\delta = \frac{3H}{\dot{\phi}}\delta\phi. \tag{8.13}$$

The correlation function that measures the mean square fluctuation between two points separated by a distance l is given by a quantum expectation value

$$\xi(l)^2 = \langle(\delta(\boldsymbol{x}, t) - \delta(\boldsymbol{x}', t))^2\rangle. \tag{8.14}$$

Substituting for δ gives

$$\xi(l)^2 = \frac{18H^2}{\dot{\phi}^2}\left(\langle\delta\phi(\boldsymbol{x}, t)^2\rangle - \langle\delta\phi(\boldsymbol{x}, t)\delta\phi(\boldsymbol{x}', t)\rangle\right). \tag{8.15}$$

It is now possible to use the limiting forms of the propagator,

$$\xi(l)^2 = \begin{cases} \dfrac{9\hbar H^4}{2\pi^2\dot{\phi}^2}\log(Hl) & \text{for } Hl \gg 1 \\[2mm] \dfrac{9\hbar H^4}{8\pi\dot{\phi}^2} & \text{for } Hl \approx 1. \end{cases} \tag{8.16}$$

Therefore the mean fluctuation size is approximately $H^2/\dot{\phi}$. Significantly, the correlation function shows no tendency to fall off with distance. This is a very special feature of massless fields and inflation. Quantum fluctuations exist during any epoch of the early universe, but generally they are minute on length scales many times the natural distance scales for elementary particle physics and they do not affect significant features of the universe today.

At some stage the quantum processes should cease to be important and the normal classical evolution of the universe should be resumed. We might suppose that this happens by the time that the length scales have become very much larger than the horizon size.

The classical density perturbation is usually represented by a random field, based upon the assumption that ensemble averages provide good approximations to volume averages of the true density field. The random field, which will still be denoted by δ, can be decomposed into comoving modes,

$$\delta(\boldsymbol{k}, t) = \int d^3x \, \delta(\boldsymbol{x}, t) \exp(i\boldsymbol{k} \cdot \boldsymbol{x}). \tag{8.17}$$

The modulus $|\delta(\boldsymbol{k}, t)|$ is what is meant by the amplitude of the density perturbation. In the classical limit we suppose that the phases $\theta(\boldsymbol{k}, t)$, where

$$\delta(\boldsymbol{k}, t) = |\delta(\boldsymbol{k}, t)| \exp(i\theta(\boldsymbol{k}, t)) \tag{8.18}$$

are completely uncorrelated.

We shall see in the next section that the amplitudes have a constant magnitude when their scale is larger than the horizon size. The amplitudes can be identified with the quantum fluctuations through the correlation function $\xi(l)^2$ of equation (8.14),

$$|\delta(\boldsymbol{k}, t)|^2 = \int d^3x \, \xi(a_c |\boldsymbol{x}|)^2 \exp(i\boldsymbol{k} \cdot \boldsymbol{x}). \tag{8.19}$$

For $Hl \gg 1$ we use a truncated form of the correlation function to simplify the integrations, but one with the correct large-distance behaviour,

$$\xi(l)^2 = \frac{9}{2\pi^2} \zeta^2 \log(Hl) \exp(-H/l) \tag{8.20}$$

with

$$\zeta = \sqrt{\hbar} \frac{H^2}{\dot{\phi}}. \tag{8.21}$$

Performing the integral for $k \ll H$ gives

$$|\delta(k, t)| = 3\zeta(k) \, k^{-3/2}. \tag{8.22}$$

Generally, H and $\dot{\phi}$ change very slowly during inflation. We should use the values when the given scale first becomes comparable in size to the physical horizon to calculate $\zeta(k)$.

8.4 The primordial spectrum

The spectrum of density perturbations, which was expressed in terms of $\zeta(k)$, depends on two functions H and $\dot{\phi}$. In the slow rolldown limit, equations

(8.8) can be used to estimate H and $\dot{\phi}$ in relation to their values at a fixed reference time t_c, chosen for convenience to be the end of the inflationary era. The inflationary era is followed by the hot Big Bang plasma phase, which is why we call this spectrum the primordial spectrum.

It proves convenient to use the field $x = G^{1/2}\phi$ as independent variable and prime for derivatives with respect to x. The equations (8.8) for slow rolldown give

$$\frac{a'}{a} = -\eta_c \frac{V}{V'} \tag{8.23}$$

where $\eta_c = (24\pi)^{1/2}$. Therefore,

$$\log\left(\frac{a(x)}{a(x_c)}\right) = -\eta_c \int_{x_c}^{x} \frac{V}{V'} dx' \tag{8.24}$$

where x_c is the value of x at the end of the inflationary phase. The mode crosses the horizon (see equation (8.2)) when x satisfies

$$\eta_c \int_{x_c}^{x} \frac{V}{V'} dx' - \log\left(\frac{V}{V_c}\right)^{1/2} = \log\left(\frac{k_c}{k}\right) \tag{8.25}$$

where $k_c = 2\pi H_c a_c$.

The amplitude for the mode k was given in terms of $\zeta = \hbar^{1/2} H^2/\dot{\phi}$. This can now be written

$$\zeta(k) = 8\pi G\hbar^{1/2}\frac{V^{3/2}}{V'} \tag{8.26}$$

with x evaluated at the solution to equation (8.25). An example of the dependence of the amplitude upon k are plotted in figure 8.4.

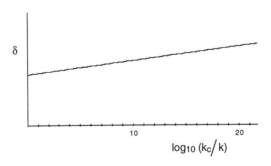

Figure 8.3 A plot showing the dependence of the fluctuation amplitude δ on the logarithm of the length scale k_c/k for chaotic inflationary models.

With the expectation that the amplitude will be only weakly dependent on k we evaluate it first for scales that cross the horizon very near to the end of the inflationary era, or $k \approx k_c$. At this time $\dot{\phi}^2 = V$, which translates to the condition

$$V'(x_c) = \eta_c V(x_c). \tag{8.27}$$

Consequently,

$$\zeta(k_c) = \tfrac{1}{3}\eta_c G\hbar^{1/2}V_c^{1/2}. \tag{8.28}$$

The slope of the spectrum is given by differentiating equations (8.25) and (8.26), and for $k \ll k_c$,

$$\frac{d\zeta}{dk} = -\frac{1}{\eta_c}\left(\frac{3}{2}\left(\frac{V'}{V}\right)^2 - \frac{V''}{V}\right)\frac{\zeta}{k}. \tag{8.29}$$

Both of these terms are small in the slow rolldown limit confirming our belief that the spectrum is nearly flat.

The spectrum of density perturbations covers the range $k < k_c$. For an idea of the value of k_c/k corresponding to a perturbation with length scale λ_L we use $\lambda_L = 2\pi a/k$, which implies

$$\frac{k_c}{k} = H_c\lambda_L. \tag{8.30}$$

The value of H_c is related to V_c by equations (8.8). It is useful to introduce the radiation temperature T_c just after inflation, which is related to V_c by energy conservation,

$$\frac{k_c}{k} = \left(\frac{H_c}{T_c}\right)T_c\lambda_L. \tag{8.31}$$

The same radiation forms the microwave background radiation today with temperature T_0. The temperature of the background radiation scales with the inverse scale factor, therefore

$$\frac{k_c}{k} = \left(\frac{H_c}{T_c}\right)T_0\lambda_0 \tag{8.32}$$

where λ_0 is the length scale of the perturbation today (excluding any non-linear effects). The ratio H_c/T_c is roughly the fourth root of V_c in Planck units, making the minimum scale for the inflationary fluctuations after expanding to the present epoch somewhat less than 1 cm.

8.5 The early evolution of density fluctuations

Once formed, we can use linear perturbation theory to follow the evolution of fluctuations through various stages of the early universe. At different times the density can be made up of various components, but each of these will be taken to be a perfect fluid.

The unperturbed spacetime can be sliced into spatially flat sections with normal vectors $\mathbf{n}$ aligned with the four-velocity of the fluid $\mathbf{u}$ (figure 8.4). The density fluctuations $\delta\rho(\mathbf{x}, t)$ will be described by

$$\delta = \frac{\delta\rho}{p + \rho} \tag{8.33}$$

where ρ and p are the total density and pressure. There are also perturbations associated with the new four-velocity $\mathbf{u}'$ and the new normal vectors $\mathbf{n}'$ of the perturbed spacetime, resulting in a first-order velocity perturbation

$$v = \mathbf{u}' - \mathbf{n}' = v^i e_i \tag{8.34}$$

The velocity perturbation belongs to the spatial sections at order v^2.

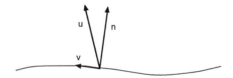

Figure 8.4 Spatial surface with normal $\mathbf{n}$ and fluid flow $\mathbf{u}$.

Useful information about the evolution of the density fluctuations can be obtained from the energy conservation equation on its own. For a perfect fluid the stress tensor has the form

$$\mathbf{T} = \rho \mathbf{g} + (\rho + p)\mathbf{u} \otimes \mathbf{u}. \tag{8.35}$$

From $\nabla \cdot \mathbf{T} = 0$ follow both the energy conservation equation

$$\dot{\rho} + (p + \rho)\theta = 0 \tag{8.36}$$

and the equation of motion

$$(\rho + p)\dot{\mathbf{u}} + \nabla_\perp p = 0. \tag{8.37}$$

The dot denotes derivative along $\mathbf{u}$ and volume expansion rates along $\mathbf{u}$ and $\mathbf{n}$ are denoted by $\theta = \nabla \cdot \mathbf{u}$ and $K = \nabla \cdot \mathbf{n}$ respectively.

The energy equation can also be written in terms of the enthalpy w,

$$\theta + \dot{w} = 0 \tag{8.38}$$

where

$$\dot{w} = \frac{\dot{\rho}}{p + \rho}. \tag{8.39}$$

Perturbing the energy equation gives

$$\delta\theta + \delta\dot{w} = 0 \tag{8.40}$$

The first term can be replaced by the divergence of equation (8.34),

$$\delta\theta = \delta K + \theta_m \tag{8.41}$$

writing θ_m for $\nabla \cdot v$. The second term can be replaced if δp is proportional to $\delta\rho$; then

$$\delta\dot{w} = \dot{\delta} \tag{8.42}$$

where δ was defined earlier in equation (8.33).

Variations of pressure and density are related by

$$\delta p = c_s^2 \delta\rho + \tau\delta s \tag{8.43}$$

with sound speed c_s and entropy density s. Entropy can be generated when more than one component of the matter content is important. This means that

$$\delta\dot{w} - \dot{\delta} = -\tau(p + \rho)^{-1}(\delta s\,\dot{w} - \dot{s}\delta). \tag{8.44}$$

The difference could be significant during the reheating stage following inflation. However, surfaces of constant time can coincide with surfaces of constant entropy, and then if $\dot{s} = 0$ the difference is very small.

In most cases we have the density fluctuation formula,

$$\delta K + \theta_m + \dot{\delta} = 0 \tag{8.45}$$

where δK is the divergence of the normals and θ_m is the divergence of the fluid flow.

This is the only equation needed to describe the density fluctuations when their scale is larger than the horizon. On this scale we may take surfaces of homogeneous expansion rate, $\delta K = 0$. Furthermore, when $k/a \ll H$, $\theta_m \sim kv/a \ll Hv \sim H\delta$ and δ is constant during this period.

The initial size of δ, set by the r.m.s. value of the fluctuation amplitude with $Hl \gg 1$, remains the value of δ up until such time as the fluctuations re-enter the physical horizon.

8.6   The later evolution of density fluctuations

The analysis of density perturbations has traditionally used the proper-time or synchronous gauge, where surfaces of constant time are separated by a uniform proper time. The metric in this gauge is

$$ds^2 = -dt^2 + a^2(\delta_{ij} + h_{ij})dx^i dx^j. \tag{8.46}$$

The metric perturbations can be decomposed into trace $h = \operatorname{tr} \mathbf{h}$ and trace-free parts h'_{ij}. Metric perturbations with $h'_{ij} = 0$ resemble a collection of Friedmann models with slightly different scale factors,

$$a(\mathbf{x}, t) = a(t)(1 + \tfrac{1}{6}h(\mathbf{x}, t)). \tag{8.47}$$

This suggests that the equation for h can be obtained from the Friedmann equation

$$\ddot{a} = -\tfrac{4}{3}\pi G(\rho + 3p)a. \tag{8.48}$$

The full set of field equations do indeed confirm this.

Variation of the Friedmann equation using equation (8.47) gives

$$\ddot{h} + 2H\dot{h} + 8\pi G(p + \rho)(\delta + 3\delta_p) = 0 \tag{8.49}$$

where $\delta_p = \delta p/(p + \rho)$.

We can also write the previous perturbation equation (8.45) in proper–time gauge,

$$\tfrac{1}{2}\dot{h} + \dot{\delta} + \theta_m = 0 \tag{8.50}$$

using the value of δK for this gauge.

One extra equation is needed. The matter equation (8.37) is already first order in small quantities, and its divergence gives

$$\dot{\theta}_m + 2H\theta_m + \frac{\dot{p}}{p + \rho}\theta_m + k^2(p + \rho)\delta_p = 0. \tag{8.51}$$

Now we have three perturbation equations to be used in conjunction with the equation of state.

The initial conditions for the density perturbations have only been given so far in the $\delta K = 0$ gauge and on scales larger than the physical horizon. We need to match these fluctuations now to the proper-time gauge. This will be done at the beginning of the era following inflation, when the universe was dominated by radiation $p = \tfrac{1}{3}\rho$.

On large scales we can neglect θ_m in equation (8.50). We also have $H^2 = 8\pi G\rho/3$, therefore equation (8.49) reads

$$\ddot{\delta} + 2H\dot{\delta} - 4H^2\delta = 0. \tag{8.52}$$

In the radiation-dominated era, $H = 1/2t$ and

$$\ddot{\delta} + \frac{1}{t}\dot{\delta} - \frac{1}{t^2}\delta = 0. \tag{8.53}$$

There is a growing solution $\delta \propto t$. This is the solution that should be matched to the primordial spectrum when the length scale crosses the physical horizon at time t_k. The amplitude in the proper-time gauge at the end of the inflationary era at time t_c is therefore

$$|\delta(k, t_c)| = 3\zeta(k)k^{-3/2}(t_c/t_k). \tag{8.54}$$

Compare equation (8.2) and the definition of k_c,

$$H(t_k)a(t_k) = k/2\pi, \tag{8.55}$$
$$H(t_c)a(t_c) = k_c/2\pi. \tag{8.56}$$

This leaves us with

$$|\delta(k, t_c)| = 3\zeta(k)k_c^{-2}k^{1/2}. \tag{8.57}$$

This is a special example of a power law spectrum,

$$|\delta(k, t_c)|^2 = A(k)k^n \tag{8.58}$$

with $n = 1$.

When the perturbation enters the horizon it becomes necessary to use the full set of equations to follow the evolution. In the radiation-dominated era the solutions for δ are damped oscillations.

During the final matter-dominated epoch, $\delta_p = 0$ and $\theta_m \to 0$ for perturbations smaller than the horizon. We can combine equation (8.50) with equation (8.49),

$$\ddot{\delta} + 2H\dot{\delta} - 4\pi G\rho\,\delta = 0. \tag{8.59}$$

Substituting the expansion law $a \propto t^{2/3}$ gives

$$\ddot{\delta} + \frac{4}{3}\frac{1}{t}\dot{\delta} - \frac{2}{3}\frac{1}{t^2}\delta = 0. \tag{8.60}$$

There is a growing mode solution proportional to the scale factor. This is the mode ultimately responsible for the existence of stars and galaxies.

Other physical effects can also be important which we have neglected here, such as interactions between the radiation and the matter content. Such effects obviously depend very much on the particular form of matter that dominates in the universe today. Due to the linearity of the perturbation equations, the evolution of the perturbations through the later stages of the early universe up to the present epoch t_0 can be represented by a transfer function $T(k)$,

$$\frac{|\delta(k, t_0)|}{|\delta(0, t_0)|} = T(k)\frac{|\delta(k, t_c)|}{|\delta(0, t_c)|}. \tag{8.61}$$

This allows the detailed development of the perturbations to be treated as a separate problem from the generation of the original spectrum.

The transfer function obtained by following the growth of the perturbations through the radiation and matter dominated eras and ignoring any other physical effects can be expressed parametrically in the form

$$T(k) = 1/[1 + ak + (bk)^{3/2} + (ck)^2] \qquad (8.62)$$

where $a = 6.4/(\Omega h^2)$, $b = 3.0/(\Omega h^2)$, $c = 1.7/(\Omega h^2)$ and $h = H_0/100 \, \mathrm{km \, s^{-1} \, MPc^{-1}}$ (Bond and Efstathiou 1984).

8.7   The cosmic microwave background

The variations in the observed temperature of the cosmic microwave background provide a direct measure of the large-scale structure from an early epoch. The microwave background probably emerged from the end of the plasma phase of the early universe, on the cosmic photosphere where matter and radiation decoupled. From its thermal origin the waves have traveled unimpeded to the present epoch.

The size of the physical horizon on the cosmic photosphere is approximatly 2 degrees. Observations on scales larger than 2 degrees should allow us to to investigate the primordial spectrum of density perturbations directly. The COBE satellite (Cosmic Microwave Background Explorer) measurements cover angular scales of at least 8 degrees.

Fluctuations in the microwave background can be caused by either scalar or tensorial gravitational perturbations. Both types of perturbation can be produced during an inflationary phase, but this section will focus on the scalar perturbations which usually predominate.

On large angular scales and for a universe with critical density, temperature fluctuations δ_T can be given in terms of a potential function Φ by

$$\delta_T = \tfrac{1}{3}\Phi. \qquad (8.63)$$

The derivation of this formula can be found in Peebles (1980). The potential Φ is equal to the scalar metric perturbation in longitudinal gauge,

$$ds^2 = -(1 + 2\Phi)dt^2 + a^2 \, \delta_{ij}(1 - 2\Phi)dx^i dx^j. \qquad (8.64)$$

The Einstein equations in this gauge imply an equation similar to Poisson's equation,

$$\widehat{\nabla}^2\Phi - 3a^2 H^2\Phi - 3a^2 H\dot{\Phi} = 4\pi G a^2 \delta\rho, \qquad (8.65)$$

where $\widehat{\nabla}$ is the flat-space gradient operator.

We may drop the spatial gradient terms in equation (8.65) for scales larger than the horizon and substitute for $H^2 = 8\pi G\rho/3$ and $\delta = \delta\rho/\rho$. We also have another perturbation equation (8.45), with

$$\delta K = -3\dot{\Phi} \tag{8.66}$$

in longitudinal gauge. This gives us two equations,

$$3\dot{\Phi} - \dot{\delta} = 0 \tag{8.67}$$

$$H^2\Phi + H\dot{\Phi} = -\tfrac{1}{2}H^2\delta. \tag{8.68}$$

An obvious solution is $\Phi = -\tfrac{1}{2}\delta = $ constant, but this coincides with the original $\delta K = 0$ gauge used in section 8.5. Therefore, using equations (8.22) and (8.63), we can write the temperature fluctuations in terms of $\zeta(k)$,

$$\delta_T = -\tfrac{1}{2}\zeta(k)\,k^{-3/2}. \tag{8.69}$$

Fluctuations in the cosmic microwave background on angular scales of a few degrees therefore still retain the amplitude of quantum correlations in the inflationary era. This is the quantity measured by the COBE satellite to be around 10^{-5}.

Observations of the cosmic microwave background will, in the future, give us an astonishing amount of information about the primordial fluctuation spectrum. This will extend from large angular scales down to angular scales smaller than the horizon scale on the cosmic photosphere. Taken in conjunction with measurements of the distribution of galaxies, this should help understand some of the physical processes taking place at very early times and extremely high temperatures.

9

Quantum cosmology

Progress towards a satisfactory theory of quantum gravity has been difficult, but some of the problems have been solved in simpler systems and may point the way to understanding quantum gravitational phenomena. Some of these systems are relevant to cosmology. The consequences are quite startling and give a suggestive picture of fundamental physics at the origin of space and time.

Throughout this chapter it will be assumed that gravity can be described by a metric which satisfies Einstein's equations in the classical limit. This presupposes that the underlying geometrical structure is a four-dimensional manifold with Lorentzian geometry. Such an assumption may well be incorrect on very small scales, but it is the natural place to begin.

In classical gravity we have a situation where time and space are not fixed by the theory and these concepts have to be introduced separately. In the language of field theory this constitutes a gauge symmetry. One consequence is that the three-dimensional geometry of space becomes the actual dynamical variable, rather than the full spacetime metric. Other components of the metric contain information about the coordinates and measure such things as the rate at which space "moves through spacetime" (DeWitt 1967a).

This leaves us with the question of what are space and time in quantum gravity. One approach would be to suppose that distances and times are the quantities measured on rulers and clocks, nothing more. An event would be localised in time through its correlation with a clock. The definition of what constitutes a clock needs to be clarified, but the clock should at least measure a classical time variable in some classical limit. We could even use a simpler approach for quantum cosmology, using the classical limit directly, ignoring material clocks and simply using the cosmological proper time set by the expansion of the universe. Of course, these approaches suppose that it is possible to recognise and isolate some kind of classical picture emerging from the quantum theory. Theories of decoherence and consistent histories (Griffiths 1984, Omnés 1988, Gell–Mann and Hartle 1990) try to do this.

The physical quantities of interest to us here can be expressed as view of the discussion of black hole evaporation. This will be important on small scales where quantum fluctuations seem large enough on dimensional grounds to form virtual black holes. It is also important in the early stages of the universe and leads to the possibility of creating the universe "ex nihilo".

In order to allow changes in topology we need to allow non-Lorentzian metrics in the path integral. An integral over purely Riemannian metrics introduces a new problem, namely the path integral diverges because the Riemannian action is not bounded below (Hawking and Israel 1979). The integral therefore has to be taken over a carefully chosen complex contour of metrics.

The correct classical limit will be obtained if the contour passes through saddle points which satisfy the classical field equations, are predominantly Lorentzian and dominate the path integral in the $\hbar \to 0$ limit. Beyond these requirements, there is not yet any universal agreement about the choice of contour. We can at least guarantee a convergent result if the contour follows paths of steepest descent through saddle points, as suggested by Hartle (Halliwell and Hartle 1990). The emphasis which this places on complex saddle points may seem strange. However, we have seen already how complex paths arise in barrier penetration, and the following simple example shows a similar situation in quantum cosmology.

Example: de Sitter space

Some of the basic ideas can be illustrated by quantising a subset of metrics which have Robertson–Walker form,

$$ds^2 = \sigma^2 \left(-N(t)^2 dt^2 + a(t)^2 d\omega_3{}^2 \right). \tag{9.1}$$

The single physical parameter is a and because of the freedom to reparametrise the time coordinate the amplitudes are independent of time, for example $\langle a_1 | a_0 \rangle$.

The Einstein–Hilbert gravitational action for gravity with a cosmological constant Λ and this metric becomes (with details later)

$$S = -\frac{1}{2} \int_0^1 \left(\frac{a\dot{a}^2}{N^2} - a + a^3 h^2 \right) N dt \tag{9.2}$$

where $h^2 = \frac{1}{3}\sigma^2 \Lambda$. In a gauge with constant N the classical solutions satisfy

$$\frac{\ddot{a}}{N^2} - h^2 a = 0 \tag{9.3}$$

and

$$\frac{\dot{a}^2}{N^2} + 1 - a^2 h^2 = 0. \tag{9.4}$$

Consider the amplitude $\langle a_1 \mid 0 \rangle$ evaluated by path integation. At order zero in $\hbar$ the amplitude is dominated by saddle points and the

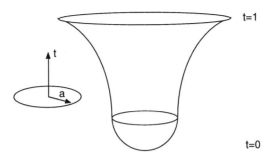

Figure 9.1 The universe emerging from nothing $(a = 0)$.

path integral measure (including gauge fixing and ghosts) enters only at order $\hbar$. There are no real solutions at all, but the complex saddle point solution is

$$a = a_1 \frac{\sin(iNht)}{\sin(iNh)} \quad \text{where} \quad \sin^2(iNh) = h^2 a_1^2. \tag{9.5}$$

For small a, N is pure imaginary and this solution resembles part of a sphere. If $a > h^{-1}$ then N has to be complex. The metric is also complex but it approaches de Sitter space for large a_1 (see figure 9.1). The contribution to the wavefunction from the action in the two cases is then

$$e^{-I} = \begin{cases} e^{\pm(1-(1-h^2a_1^2)^{3/2})/3h^2} & ha_1 < 1 \\ e^{\pm 1/3h^2} e^{\pm i(h^2a_1^2-1)^{3/2}/3h^2} & ha_1 > 1. \end{cases} \tag{9.6}$$

The two signs correspond to the choice of sign for N and represent two of the four saddle point solutions. The other two saddle points represent the freedom to place the spatial sections in the expanding or contracting regions of spacetime. In figures 9.4 and 9.5 we see that the wavefunction oscillates in the Lorentzian region and behaves exponentially in the Riemannian region.

9.1 Decoherence

The interpretational framework is an important issue in quantum cosmology. Predictions about, or at least consistency with, the values of astronomical

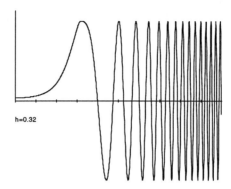

h=0.32

Figure 9.2   Growing wavefunction for de Sitter space at $O(1)$ in $\hbar$.

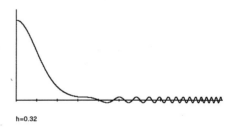

h=0.32

Figure 9.3   Decaying wavefunction for de Sitter space at $O(1)$ in $\hbar$.

parameters must somehow be extracted from the quantum formalism. This has to take into account several features that are particularly cosmological. We have already seen that time in the theory has an intrinsic character, rather than appearing as an explicit label of spacetime events. Also, the observer is a small part of the whole system which is being quantised. The universe may start out being dominated by quantum uncertainty, but eventually we know that it becomes very large and completely classical.

For many problems in quantum cosmology a pragmatic approach to the interpretation of the wavefunction has been used; configurations where the wavefunction has a large amplitude are taken as predictions of these configurations with a high probability. The wave function implies that certain values of the variables are correlated, but nothing is said in this approach about the interdependence of different measurements.

Another feature missing from this approach is any information about dynamics, information that would be necessary to reconstruct the history of the universe from its beginning. Again there is a simplistic approach, which says that when the wavefunction has a semi-classical form then this is taken to be a prediction of the underlying classical trajectory. This can be made a little more precise by introducing the Wigner function,

$$W(x, p) = \int_{-\infty}^{\infty} e^{2ipy} \psi^*(x + y)\psi(x - y)\, dy. \tag{9.7}$$

The Wigner function of a semi-classical state is peaked along the classical phase-space trajectory. Since the expectation values of operators $A(x, p)$ can be found by weighting them with the Wigner function, this picks out nicely the preferred trajectory. Unfortunately, the Wigner function is not always positive and cannot be interpreted as a probability measure.

Some progress towards a better understanding of how the classical world arises from a quantum wavefunction can be achieved using the theory of decoherence. In a decohered measurement, quantum interference effects are suppressed by the averaging out of microscopic variations not distinguished by the associated observable. For example, if the configuration variables of the system are divided in some way into macroscopic variables $\mathcal{M}$ and microscopic variables $\mathcal{Q}$, then the expectation value for any macroscopic operator A sensitive to $x \in \mathcal{M}$ but insensitive to $q \in \mathcal{Q}$ can be written

$$\langle A \rangle = \int_{\mathcal{Q} \cup \mathcal{M}} d\mu\, \psi^*(x, q, t)\psi(x', q, t)A(x, x'). \tag{9.8}$$

Therefore, for measurements on the macroscopic features of the system we may as well use the *reduced* density matrix,

$$\rho_R(x', x, t) = \int_{\mathcal{Q}} d\mu \psi^*(x, q, t)\psi(x', q, t) \tag{9.9}$$

as an integration kernel.

An interesting phenomenon now takes place. In many systems, because the phase of the wavefunction is very sensitive to variations in x, the off-diagonal elements of ρ_R become small in comparison with the diagonal elements. The expectation values become

$$\langle A \rangle = \int_{\mathcal{M}} d\mu\, \rho_R(x, x, t)A(x, x) \tag{9.10}$$

the same as the average value of a (classical) random function with probability distribution $\rho_R(x, x, t)$.

Example: Decoherence by coupled oscillators.

Consider a variable x and a collection of oscillators q_n with frequencies $\omega_n(x)$, all in their ground states. We shall see later that this system pops up in quantum cosmology, with x the scale factor and q_n representing photons and gravitons. Then,

$$\psi(x,q) = \psi_0(x) \prod_n \left(\frac{\omega_n(x)}{\pi} \right)^{1/4} e^{-\omega_n(x)q_n^2/2}. \qquad (9.11)$$

If only measurements of x are made, then the other oscillator modes can be eliminated,

$$\rho_R(x',x) = \int d\mu(q)\psi_0^*(x')\psi^*(x',q)\psi_0(x)\psi(x,q). \qquad (9.12)$$

Hence

$$\rho_R(x',x) = \psi_0^*(x')\psi_0(x) \prod_\omega \sqrt{2} \left(\frac{\omega(x)}{\omega(x')} + \frac{\omega(x')}{\omega(x)} \right)^{-1/2}. \qquad (9.13)$$

Alternatively,

$$\rho_R(x',x) = \psi_0^*(x')\psi_0(x) \prod_\omega \exp\left(-\tfrac{1}{2}\log\left(1 + \frac{(\omega(x)-\omega(x'))^2}{2\omega(x)\omega(x')} \right) \right). $$
$$(9.14)$$

Given that $\omega_n(x) \sim nx^{3/2}$ for large n in the cosmological case, it follows that the reduced density matrix vanishes when $x' \neq x$.

Decoherence of the density matrix shows how quantum interference effects can be suppressed for measurements at a single time but we still need to combine this with time evolution to form a complete picture of the classically evolving system. In path integral form the transition amplitude for the reduced density matrix between times $t=0$ and $t=t_f$ would be

$$\rho(x_f', x_f; t_f) = \int_0^{t_f} d\mu[x', q', x, q]\rho(x_0', q_0'; x_0, q_0; 0)e^{iS[x,q]/\hbar}e^{-iS[x',q']/\hbar}$$
$$(9.15)$$

where x_f is given, but q_f is integrated out. Interactions between x and q are described by the action $S[x,q]$,

$$S[x,q] = S[x] + iS_I[x,q]. \qquad (9.16)$$

If the initial density matrix factorises, we can separate the integrals over q from the integrals over x. The integrals over q define a functional $\mathcal{F}[x,x']$, the Feynman–Vernon influence functional (Feynman and Hibbs 1965),

$$\mathcal{F}[x,x'] = \int_0^t d\mu[q',q]\rho(q_0';q_0;0)\mathcal{P}'(q_f,t_f)\mathcal{P}(q_f,t_f) \qquad (9.17)$$
$$\times \exp\left(iS_I[x,q] - iS_I[x',q'] \right)/\hbar. \qquad (9.18)$$

We can also re-express the influence functional as an influence action $S_{IF}[x, x'] = -i \log \mathcal{F}[x, x']$. The influence action has both real and imaginary parts, $S_{IF} = \mathcal{R} + i\mathcal{I}$. It follows from the definition of the influence functional that $\mathcal{I}[x, x'] = \mathcal{I}[x', x]$ and $\mathcal{I}[x, x] = 0$. Therefore the leading term in the functional expansion of $\mathcal{I}$ for small $x - x'$ is generally quadratic, and the influence functional has a peak which damps out pairs of trajectories with $x \neq x'$.

The probabilities at the final time t_f are obtained by setting $x'_f = x_f$ in the Feynman–Vernon influence functional. The resulting quantity introduces the decoherence functional,

$$\mathcal{D}[x, x'] = \mathcal{F}[x, x']|_{x'_f = x_f} . \tag{9.19}$$

The decoherence functional brings us to a rather different picture of decoherence where the emphasis is on the time evolution histories rather than the density matrix. This is the theory of consistent histories. The path integral version of consistent histories begins by choosing conditions which partition the trajectories into separate sets $\mathcal{C}$. These conditions may be ranges of values Δ^t through which the trajectory passes at fixed times t, for example. In the system described above Δ^t would include all values of q and a single value of x for all $0 < t < t_f$. Each set of trajectories defines a history.

The amplitude for a given history is given by a path integral

$$\langle \mathcal{C} \rangle = \int_0^{t_f} d\mu[x] P(\mathcal{C}) e^{iS(x)} \tag{9.20}$$

where $P(\mathcal{C}) = 1$ if $x(t) \in \Delta^t$, and zero otherwise. The decoherence functional is defined by pairs of histories,

$$D(\mathcal{C}_1, \mathcal{C}_2) = \int \langle \mathcal{C}_1 \rangle^* \langle \mathcal{C}_2 \rangle \rho(x'_0, x_0) dx'_0 dx_0 \tag{9.21}$$

where both trajectories meet at x_f and t_f. Histories are said to be consistent when $D(\mathcal{C}_1, \mathcal{C}_2)$ vanishes, or is very close to zero, on different histories $\mathcal{C}_1$ and $\mathcal{C}_2$. The diagonal elements of the decoherence functional for consistent histories are interpreted as the probabilities of the decohered histories. The decoherence functional has two important properties that suggest this interpretation: namely

1. $\mathcal{D}(\mathcal{C}, \mathcal{C}) \geq 0$.
2. $\mathcal{D}(\mathcal{C}_1 \cup \mathcal{C}_2, \mathcal{C}_1 \cup \mathcal{C}_2) = \mathcal{D}(\mathcal{C}_1, \mathcal{C}_1) + \mathcal{D}(\mathcal{C}_2, \mathcal{C}_2) + 2\Re \mathcal{D}(\mathcal{C}_1, \mathcal{C}_2)$.

The second of these properties implies that probabilities of different consistent histories can be summed to get the probability of the combined history. This is a distinctive feature of consistent histories and represents a loss of quantum interference effects. It is possible to include the most probable consistent histories in a given theory as predictions of that theory and regard them in tha same way as predictions from classical physics.

9.2 The initial state

In classical cosmology restrictions are placed upon the initial form of the metric, expansion and density fields that select a unique solution to the field equations that can represent the universe. The corresponding problem in quantum theory is the subject of this section, the selection of a wavefunction or possibly a density matrix to represent the universe.

In quantum cosmology there is an extra step of interpretation involving the selection of a classical universe from the quantum picture represented by the wavefunction. Even full knowledge of the wavefunction itself may only be partially helpful in relating the predictions to the observations.

For classical cosmology we are guided by the Copernican principle, that we do not occupy a privileged place in the universe (Weinberg 1972). When combined with the observed isotropy of the universe, this implies that the universe is also homogeneous. The Copernican principle seems to express a view that occupying a privileged place in the universe is a very unlikely event. The quantum analogue would be to say that the universe which we see is a likely (high-probability) outcome of an underlying quantum theory. Therefore, in selecting a wavefunction for the universe, we should rule out any candidate in which homogeneity and isotropy have low-probability.

One quantum state in particular has been the centre of attention, suggested by Hartle and Hawking (1983). This state is defined by

$$\psi(\boldsymbol{h}, \phi) = \int d\mu[\mathbf{g}, \phi] e^{-I} \tag{9.22}$$

where the metrics included in the path integral are all closed Riemannian with boundary $\boldsymbol{h}$. Evaluation of the integral usually requires some form of analytic continuation, for example integrating along contours of steepest descent through saddle point solutions.

Simplified models can be constructed by ignoring most of the degrees of freedom. As in the example which we looked at earlier, the path integral is dominated by complex metrics which are predominantly Riemannian or Lorentzian when the universe is very small or very large respectively. In general, the wavefunction can have contributions from many such metrics and be a superposition of semi-classical components. Thus

$$\psi = \sum A_n e^{iS_n} \tag{9.23}$$

It is tempting to interpret this state as a probability distribution $|A_n|^2$ on an ensemble of classical cosmologies. This can be checked in particular models where it should be possible to see whether these semiclassical histories decohere in the ways described in the previous section.

The amplitudes A_n contain factors of e^{-I} from the action of the Riemannian part of the path. This Riemannian region has much in common with the instanton solutions that arise in the theory of quantum tunnelling, and it can be viewed as a gravitational instanton which gave birth to our universe. In this sense, the universe may have arisen as a quantum tunnelling event.

The choice of saddle points in the integration is important and affects the sign of the Riemannian action. In the earlier example, the negative sign corresponds to an alternative quantum state proposed by Vilenkin and the positive sign to the original result of Hartle and Hawking. If we are restricted to path integral definitions of states, then the second choice is the more acceptable one because it is the one which gives the correct sign for coupling matter to gravity.

9.3   The Wheeler–DeWitt equation

It is possible to replace the path integral description of gravitational transition amplitudes with a set of functional differential equations. The symmetry under changes in coordinates plays an important role in these equations, most notably in the absence of the time derivatives which are present in the ordinary Schrödinger equation.

The first step in deriving the equations is to decompose the spacetime into three-dimensional hypersurfaces Σ_t. The phase space consists of 3-metrics h_{ij} and various matter fields ϕ on Σ_t, together with their conjugate momenta p^{ij} and π. The result of this decomposition is given in appendix B,

$$S = \int \left(\dot{h}_{ij} p^{ij} + \dot{\phi}\pi - N\mathcal{H} - N^i \mathcal{H}_i \right) d^4x. \qquad (9.24)$$

There are no momenta conjugate to the remaining metric components N and N^i. The 'superhamiltonian' $\mathcal{H}$ can be decomposed into gravitational and matter parts,

$$\mathcal{H}(h, p, \phi, \pi) = \mathcal{H}_g(h, p) + \mathcal{H}_m(h, \phi, \pi). \qquad (9.25)$$

From the purely Einstein action, decomposition gives

$$\mathcal{H} = \tfrac{1}{2} G_{ijkl} p^{ij} p^{kl} - r\sqrt{h}/(2\kappa^2) \qquad (9.26)$$

$$\mathcal{H}_i = -2h_{ik} p^{kj}{}_{|j} \qquad (9.27)$$

where $\kappa^2 = 8\pi G$, the vertical bar denotes the h-covariant derivative with Ricci scalar r and G_{ijkl} is the DeWitt metric,

$$G_{ijkl} = \kappa^2 \left(h_{ij} h_{kl} + h_{il} h_{jk} - h_{ij} h_{kl} \right) / \sqrt{h}. \tag{9.28}$$

When taking covariant derivatives of the supermomentum it is important to remember that p^{ij} differs from a pure tensor by a factor $h^{1/2}$.

The lapse and shift functions N and N^i act as Lagrange multipliers for the constraint equations $\mathcal{H} = 0$ and $\mathcal{H}_i = 0$. These are a subset of Einstein's equations. The other Einstein equations follow from the variation of the action with respect to h and p. Their classical solutions are represented by trajectories in configuration space, with different choices of lapse and shift functions leading to different trajectories to represent the same spacetime.

In the quantum theory states can be represented by wave functionals $\psi(h, \phi)$. Dirac's approach to quantisation imposes the constraints by operator equations

$$\mathcal{H}\psi \;=\; 0 \tag{9.29}$$

$$\mathcal{H}_i\psi \;=\; 0 \tag{9.30}$$

with p^{ij} replaced by $-i\partial/\partial h_{ij}$. These are similar to the wave equations that arose for the relativistic particle in chapter two. The first equation is usually known as the Wheeler–DeWitt equation.

Because of the vanishing of the superhamiltonian $\mathcal{H}$ it is not possible to introduce a $\partial\psi/\partial t$ term on the right-hand side of the Wheeler–DeWitt equation. This is an important difference from the usual Schrödinger equation, but similar to the wave equation of the relativistic particle that was discussed in chapter 2. The Wheeler–DeWitt equation is implicitly a dynamical equation because the signature of the superspace metric implies that $\mathcal{H}$ is a (pseudo-)hyperbolic operator. This hyperbolicity is directly related to the lack of positive definiteness of the superhamiltonian in the classical theory.

The fact that ψ does not depend on time is simply an expression of general covariance, time being a coordinate label. Physical questions about time development can be addressed by choosing some degrees of freedom to form clock subsystems against which the time development of the remaining system can be measured.

There is a factor ordering problem associated with the form of the momentum term in the Wheeler–DeWitt equation which can be partially resolved by imposing covariance under changes of h_{ij}. The simplest possibility would be to use the superspace Laplacian,

$$\nabla^2 \simeq |G|^{-1/2} \frac{\delta}{\delta h_{ij}} G_{ijkl} |G|^{1/2} \frac{\delta}{\delta h_{kl}}. \tag{9.31}$$

This requires some suitable definition of the infinite determinant $|G|$. The

Wheeler–DeWitt equation would then be

$$\mathcal{H}_g = -\tfrac{1}{2}\nabla^2 + \tfrac{1}{2}U \tag{9.32}$$

where $U = r\sqrt{h}/\kappa^2$.

Factor ordering is also an issue in the commutator algebra of the constraints. To write down the constraint algebra, equations (9.26) and (9.27) should be replaced by distributions,

$$(\mathcal{H}, f) = \int_{\mathcal{M}} d\mu\, \mathcal{H}f \tag{9.33}$$

$$(\mathcal{H}, v) = \int_{\mathcal{M}} d\mu\, \mathcal{H}_i v^i. \tag{9.34}$$

The constraints are then

$$[(\mathcal{H}, v_1), (\mathcal{H}, v_2)]_{PB} = (\mathcal{H}, [v_1, v_2]) \tag{9.35}$$
$$[(\mathcal{H}, v), (\mathcal{H}, f)]_{PB} = (\mathcal{H}, \mathcal{L}_v f) \tag{9.36}$$
$$[(\mathcal{H}, f_1), (\mathcal{H}, f_2)]_{PB} = (\mathcal{H}, \boldsymbol{J}(f_1, f_2)). \tag{9.37}$$

In the last term, $\boldsymbol{J}(f_1, f_2)$ is the vector dual to $f_1 df_2 - f_2 df_1$.

Consistency requires that commutators also vanish as operator equations. This can easily lead to factor ordering problems due to the non-linearity of the constraints. Another difficulty is the presence of the metric in $\boldsymbol{J}$ which implies that the constraint algebra is not even a Lie algebra. Furthermore, nothing has yet been done to remove divergences. The regularisation procedure that removes divergences may well introduce extra terms on the right-hand side of the commutator relations, i.e. anomalies, which would render the theory inconsistent.

9.4 Minisuperspace

The idea of minisuperspace is to replace the three-dimensional metric h_{ij} by a homogeneous metric, reducing a problem with infinitely many degrees of freedom to one with a finite number. It need hardly be said that, as an approximation, this one is extremely crude. On the other hand, it gives us a chance to see how some of the preceding ideas work in practice.

Consider a metric and a scalar field which are

$$ds^2 = \sigma^2 \left(-N(t)^2 dt^2 + a(t)^2 d\hat{x}.d\hat{x}\right) \tag{9.38}$$

$$\phi = q(t)/\sqrt{2}\pi\sigma \tag{9.39}$$

where $\sigma^2 = \kappa^2/12\pi^2$ is a normalising constant and $d\hat{x}.d\hat{x}$ is the metric on a unit three-sphere. Minisuperspace is the configuration space with points (a, q).

The Einstein action

$$S_g = \frac{1}{2\kappa^2} \int_M Rd\mu + \frac{1}{\kappa^2} \int_{\partial M} kd\mu \qquad (9.40)$$

and the matter action

$$S_\phi = - \int_M \left(\tfrac{1}{2} h^{ab} \nabla_a \phi \nabla_b \phi + V(\phi) \right) d\mu. \qquad (9.41)$$

The curvature and extrinsic curvature scalars for the metric of this form become

$$R = 6\sigma^{-2} \left(a^{-1} a'' + a^{-2} \left(1 + (a')^2 \right) \right) \qquad k = a^{-1} a' \qquad (9.42)$$

where $a' = N^{-1} \dot{a}$. The total action can be written, after integrating over a spatial volume $2\pi^2$, as

$$S = \int Ldt \qquad (9.43)$$

where the Lagrangian is

$$L = -\tfrac{1}{2} a N^{-1} \dot{a}^2 + \tfrac{1}{2} a^3 N^{-1} \dot{q}^2 + \tfrac{1}{2} Na - 2Na^3 \pi^2 \sigma^4 V. \qquad (9.44)$$

Proceeding to the Hamiltonian, we have the conjugate momenta to replace the time derivatives,

$$\mathbf{p} = N^{-1} \left(-a\dot{a}, a^3 \dot{q} \right). \qquad (9.45)$$

The natural metric on minisuperspace is defined by the kinetic term in the Lagrangian,

$$d\mathcal{S}^2 = -ada^2 + a^3 dq^2. \qquad (9.46)$$

This metric appears in the conjugate momenta, which should therefore be regarded as cotangent vectors. The inverse metric appears in the Hamiltonian,

$$H = \tfrac{1}{2} \mathbf{p} \cdot \mathbf{p} + \tfrac{1}{2} U(a, q) \qquad (9.47)$$

where

$$U(a, q) = 4a^3 \pi^2 \sigma^4 V(q) - a. \qquad (9.48)$$

Quantisation presents us with the problem of ordering the factor of $\mathbf{p}$ with respect to the minisuperspace metric. One choice is particularly appealing, using the $\mathcal{S}$ connection ∇, because this makes the Wheeler–DeWitt equation covariant under changes of parameters,

$$(-\nabla \cdot \nabla + U) \psi = 0. \qquad (9.49)$$

This is the starting point for many quantum cosmological models. For inflationary models it should be arranged that the potential acts as if it were a cosmological constant for some regions of phase space. This ensures that the Wheeler–DeWitt equation has solutions which behave as in the de Sitter example for a range of field values. At large values of the scale factor the wavefunction can be approximated by a sum of semiclassical terms,

$$\psi = \sum_n A_n e^{iS_n(a,q)} \tag{9.50}$$

where the functions S_n satisfy the Hamilton–Jacobi equations for the system and can be used to recover classical trajectories, here corresponding to classical cosmological models.

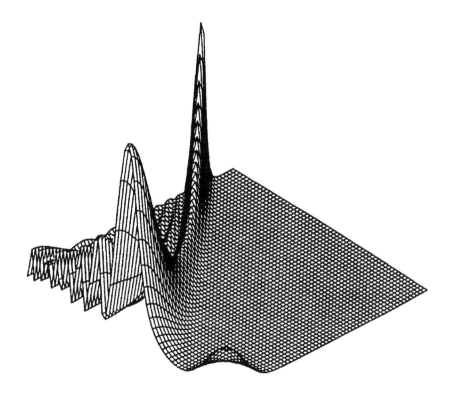

Figure 9.4 The probability density for a Higgs model universe, with coordinate a along the left edge and $\tilde{q} = aq$ along the right edge.

Example: Higgs models.

It is convenient here to rescale the scalar field q by the scale factor a to $\tilde{q} = aq$ and parametrise minisuperspace by $(a, \tilde{q})$. The numerical solution shown in figure 9.4 uses a Higgs potential

$$V(\phi) = \tfrac{1}{12} R\phi^2 - \tfrac{1}{2}\mu^2\phi^2 + \tfrac{1}{4}\lambda\phi^4. \tag{9.51}$$

In cases like this it is possible to integrate the Wheeler–DeWitt equation with boundary conditions at $a = 0$ set, in this example, by the path integral for the Hartle-Hawking state.

The classical system would be inflating at the symmetric minimum of the scalar field potential, $\phi = 0$ or $\tilde{q} = 0$, due to the vacuum energy. It would be expanding more slowly near to the global minimum $\tilde{q} = a\mu^2/\lambda$ as the scalar field oscillations die away.

The wavefunction is concentrated near to the symmetric minimum for small values of the scale factor, where it resembles the wavefunction of de Sitter space. At larger values of the scale factor the wavefunction is at its largest near to the global minimum, indicating tunnelling from the symmetric minimum. These correlations between the scale factor and the matter field allows the scale factor to be used like a cosmological clock with which to follow the evolving matter field. However, this is limited by the fact that at later times the wave–packet contains not just one but many semi–classical universes which have undergone inflationary phases of various lengths.

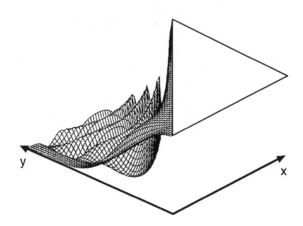

Figure 9.5   The wavefunction for a chaotic model universe.

Example: Chaotic models

For a chaotic model with potential $V(\phi) = \frac{1}{2}m^2\phi^2$ it is convenient to introduce coordinates $x = a \sinh q$ and $y = a \cosh q$ for minisuperspace. A version of the wave function starting from the Hartle–Hawking boundry data is shown in figure 9.5. In the semiclassical limit classical trajectories are perpendicular to the wave crests. In this example it is very evident that there are many classical trajectories, each of which can be parametrised by its position along an individual wave crest. These trajectories have all undergone a period of inflation. They represent different possibilities for the initial value of the scalar field. Close to $a = 0$ the wave function is uniform with respect to q. At larger values of a the wave function is a superposition of semi–classical wave functions, each of which represents a different classical universe.

The actual wave function grows exponentially in y for $x \approx 0$. This growth has no classical counterpart as a Lorentzian geometry and therefore this feature has been removed from the wave function shown in figure 9.5 with an exponential damping factor.

9.5   Reduction to minisuperspace

In classical relativity a great deal has been learnt from analysing solutions to the Einstein equations with a high degree of symmetry. The analysis of minisuperspace models has likewise given an insight into many interesting areas of quantum cosmology. The important difference here is the fact that even symmetric quantum amplitudes will be subject to effects from fluctuations involving asymmetric metrics. Even pure gauge fluctuations should be taken into account. Even worse, the total effect of all these fluctuations introduces divergences that have to be removed. It seems impossible that the minisuperspace approach can have any relevance.

The natural way of improving on the minisuperspace approach is to begin with metrics that are close to minisuperspace models. In this way we might check the consistency of the minisuperspace truncation. These perturbed minisuperspace models can also be used to describe spatial fluctuations.

We could proceed to solve the Hamiltonian constraint perturbatively. However, there is an alternative proceedure which involves fixing the gauge freedom associated with transformations of the metric perturbations γ and using the gauge-fixed covariant action of chapter 4. In this case it is possible to use either an operator equation or a path integral to evaluate the wave function. The path integral formulation has advantages over the canonical approach. Perturbation theory is explicitly covariant and the divergences are of recognisable types that should give us some confidence in the results.

Since inflationary models are particularly interesting, it is desirable to include the scalar matter field that is associated with inflation. The full complement of fields is therefore the metric perturbation γ_{ab}, the gravity ghost c_a together with its conjugate, and the scalar field perturbation Φ. For the minisuperspace background we can take the same model as in the previous section, with coordinates (a, q).

The Lagrangian density for gravity, with gauge fixing and ghosts, was given in chapter 4. The symmetry of the background metric implies that it will be useful to decompose the fields into separate space and time components,

$$\gamma_{ab} = 2\kappa\sigma^2 \begin{pmatrix} \sqrt{2}h^N + \sqrt{6}h^L & \frac{1}{\sqrt{2}}h_i^V \\ \frac{1}{\sqrt{2}}h_j^V & h_{ij}^T + \frac{1}{\sqrt{6}}h^L\eta_{ij} \end{pmatrix} \qquad c_a = \begin{pmatrix} \beta \\ \alpha_i \end{pmatrix}. \qquad (9.52)$$

Numerical coefficients have been chosen to simplify the normalisation of the kinetic terms in the Lagrangian density.

By fixing the gauge we have removed the need for most of the gauge constraints. The residual BRST symmetry implies a constraint, as does the residual symmetry under changes in the time coordinate. In the operator approach to quantisation, both constraints are imposed by the vanishing of operators acting on states of the system. In the path integral approach the BRST constraint can be imposed by boundary conditions on the fields in the path integral.

Transformations of the modes under BRST are denoted by s,

$$s\,\gamma_{ab} = c_{a;b} + c_{b;a} \qquad (9.53)$$
$$s\,\Phi = c^a\Phi_{;a}. \qquad (9.54)$$

These transformations should be decomposed into space and time components. They can be realised in the quantum theory by commutators

$$is\,h^X = [h^X, \mathbf{Q}] \qquad (9.55)$$

with the BRST charge

$$\mathbf{Q} = \sum_{X=T\ldots} \pi^X \,\widehat{s}\, h^X \qquad (9.56)$$

where $\widehat{s}$ denotes dropping any time derivatives from s, these being already included in the π^Xs.

The full expression for $\mathbf{Q}$ will not be quoted here, but it can be made to vanish by imposing the following conditions:

$$\alpha_a = \beta = 0 \qquad (9.57)$$
$$h_{ab}^{TT}, \ \Phi \ \text{given} \qquad (9.58)$$
$$h^L = h_{ab}^T{}^{|b} = 0 \qquad (9.59)$$
$$\pi_a^V + 4Hh_a^V - h_{|a}^N = 0 \qquad (9.60)$$
$$\pi^N + 6Hh^N + h_a^V{}^{|a} - \sqrt{2}\kappa\dot{q}\Phi = 0 \qquad (9.61)$$

where h_{ab}^{TT} is the transverse-traceless component of h_{ab}^T.

Vanishing of one of the variables q_n^X in the operator sense means that $q_n^X \psi = 0$. For the ghost field, this has the implication that the ghosts' contribution to the Hamiltonian is zero. In the path integral, conditions involving π_n^X should be translated into conditions on the time derivative of the field. These give Robin boundary conditions similar in form to those mentioned in appendix A.

9.6   The semiclassical limit

The symmetry of the background metric enables us to reduce the spatial dependences of the fields to mode sums. Harmonic analysis on the hypersphere is described in appendix D. There are two sequences of vector and three sequences of tensor modes,

$$h_{ij}^T = a^2 \sum_n q_n^R Q_{nij} + a^2 \sum_n q_n^S P_{nij} + a^2 \sum_n q_n^T G_{nij} \qquad (9.62)$$

$$h_i^V = a \sum_n q_n^U Q_{ni} + a \sum_n q_n^V P_{ni}. \qquad (9.63)$$

The scalar coefficients are listed in table 9.1. Similar expansions can be used for the ghosts, with coefficients q_n^A, q_n^B and q_n^C.

Table 9.1   Oscillator frequencies $\omega^2 = c_1(\dot{a}/a)^2 + c_2 a^{-2}$ for the perturbative modes. Both $\dot{H}$ and $\dot{q}$ are assumed negligible.

Mode	Field	c_1	c_2
T	h^T	0	$\lambda^S - 5$
S	h^T	0	$\lambda^P - 1$
R	h^T	0	$\lambda^G + 1$
U	h^V	1	$\lambda^S + 1$
V	h^V	1	$\lambda^P + 1$
L	h^L	1	$\lambda^S - 3$
N	h^N	10	$\lambda^S - 1$
P	ϕ	0	λ^S

Substitution of these expressions into the action is straightforward but lengthy. The details will therefore be omitted, but the result is a Lagrangian

$$L = L_0 + L_1 \qquad (9.64)$$

where L_0 is the minisuperspace Lagrangian (9.44) and

$$L_1 = N \sum_{n,X=T\dots} \tfrac{1}{2} a^3 \left((\dot{q}_n^X)^2 - (\omega_n^X)^2 (q_n^X)^2 \right). \qquad (9.65)$$

The frequencies are given in table 9.1.

There is one constraint from the remaining symmetry under time reparametrisation. This constraint fixes the Hamiltonian,

$$H = H_0 + H_1 \qquad (9.66)$$

where H_0 is the minisuperspace Hamiltonian (9.47), and

$$H_1 = \sum_{n,X=T\dots} \tfrac{1}{2} \left(a^{-3} (p_n^X)^2 + a^3 (\omega_n^X)^2 (q_n^X)^2 \right). \qquad (9.67)$$

Breaking the symmetry under coordinate transformations means that the Hamiltonian no longer vanishes locally.

The Dirac quantisation procedure implies that the Hamiltonian constraint should be converted into an operator and applied to the wavefunction,

$$(H_0 + H_1)\psi = 0. \qquad (9.68)$$

Treating the BRST constraint in the same way implies that we can drop the ghost fields and impose the conditions (9.57)–(9.61).

Approximate solutions can be constructed that represent the semiclassical limit,

$$\psi(a, q, q_n^X) = \psi_0(a, q)\psi_1(a, q, q^X) \qquad (9.69)$$

with $H_0\Phi_0 = 0$. In the WKB approximation, given classical solutions $a_c(t)$ and $q_c(t)$, and action $S_c(a, q)$,

$$\Psi_0 = A(a, q) \exp(iS_c/\hbar). \qquad (9.70)$$

Using ∇ for the minisuperspace gradient operator, the classical momentum $\mathbf{p} = \nabla S_c$. Substituting the approximate wavefunction into the constraint gives

$$\psi_0(\hbar \nabla S_c \cdot \nabla \psi_1 + H_1 \psi_1) = O(\hbar^2). \qquad (9.71)$$

Time can be defined in terms of the classical solution trajectories, then

$$\frac{d}{dt} = \mathbf{p} \cdot \nabla = \nabla S_c \cdot \nabla. \qquad (9.72)$$

Therefore, to leading order in $\hbar$, we arrive at Schrödinger's equation on the Robertson–Walker spacetime

$$-i\hbar \frac{d}{dt}\psi_1 = H_1 \psi_1. \qquad (9.73)$$

Given the form of H_1, the solutions can be expanded in harmonic oscillator wavefunctions Ψ_N,

$$\psi_1(q^X, t) = \sum_N c_N(t) \Psi_N(q^X, t). \tag{9.74}$$

Once the universe gets larger than the Planck size, $a \gg \kappa$, then the occupation numbers $c_N(t)$ are effectively frozen (provided that $\dot{q}$ remains small).

The alternative way to quantise is to use a path integral. Boundary conditions are specified in terms of values of the scale factor a, the scalar field q and the perturbation fields q^X. For definiteness, we can choose the Hartle–Hawking state with $a = 0$ initially and take the spacetime to be Riemannian.

In order to construct the $\hbar$ expansion we expand the fields about background fields, replacing q^X by $Q^X + q^X$. The background part Q^X can be given the given boundary data, whilst the quantum part q^X vanishes on the boundary.

The integral over the quantum fields precisely parallels the definition of the effective action for gauge-fixed gravity. The result depends only on the final values of a and q, and will be denoted by $\Gamma_c(a, q)$. The backgrounds contribute to the classical oscillator action $S_c(q_n^X)$. Consequently,

$$\Psi(a, q, q^X) = \exp(-\Gamma_c(a, q)/\hbar) \prod_{n, X=L,T,P} \exp(-\tfrac{1}{2} a^6 \omega_n^X q_n^{X\,2}) \tag{9.75}$$

where

$$\Gamma_c = S_c + \tfrac{1}{2}\hbar \log \det \Delta_L + \tfrac{1}{2}\hbar \log \det \Delta_\phi - \hbar \log \det \Delta_{gh} + O(\hbar^2). \tag{9.76}$$

The path integral therefore produces results very similar to the operator approach. It handles the back-reaction of the oscillators on the spacetime more accurately, but it has only been defined for Riemannian backgrounds.

The determinants are evaluated on Riemannian spaces with boundaries. The Riemannian spaces represent quantum tunnelling. In the simplest inflationary models, where the energy density is dominated by constant vacuum energy, the background spaces are approximately spherical caps, ranging in size from nothing up to a hemisphere. The value of the effective action for the hemisphere fixes the modulus of the wavefunction, and therefore the probability. These values depend on the vacuum energy and through this the initial matter field configuration.

The boundary of the hemisphere marks the beginning of the Lorentzian evolution. We can take the wavefunction there to determine the initial conditions for the wavefunction of quantum fluctuations in the inflationary era. These fluctuations are the same as the fluctuations that where discussed in connection with inflation in chapter 7.

9.7   Outlook

This discussion of the origin of our universe as a quantum event can only give a flavour of what the future developments in fundamental physics will reveal. The whole picture presented here has been based upon Einstein's theory of gravity even though the divergences in this theory present severe problems.

Alternative ways of quantising gravity, especially the use of non–local variables, have hardly been touched upon. Neither have alternative theories of gravity, such as superstring theories. Many of the developments in these, as well as other, areas use the same techniques that have been presented in this book. These alternatives can provide a very different insight into the important problems, but they can also be seen as examples of quantum field theories.

Despite the necessary warnings, working in this area of fundamental physics results in a feeling of excitement and a sense that something fundamental about the nature of time is being learned.

Appendix A

Heat kernel expansions

The small-time expansion of a heat kernel provides us with a great deal of information about operators and their eigenvalues. This expansion was substantially developed in the 1960s, and more of the mathematical details can be found in the original work and reviews by DeWitt (1964), McKean and Singer (1967) and Gilkey (1984).

The heat kernel of the operator Δ is defined by

$$K(\mathbf{x}, \mathbf{x}'; t) = \sum_n u_n(\mathbf{x}) u_n(\mathbf{x}') e^{-\lambda_n t} \qquad (A.1)$$

where $u_n(x)$ are normalised eigenfunctions with eigenvalues λ_n. The two separate spacetime points are somewhat superfluous and an integrated form of the heat kernel is often used,

$$K(f, t) = \int d\mu(x) f(\mathbf{x}) K(\mathbf{x}, \mathbf{x}; t). \qquad (A.2)$$

We consider operators of the form

$$- \mathbf{D}^2 + \mathbf{X} \qquad (A.3)$$

with a gauge-covariant derivative $\mathbf{D} = \nabla + \mathbf{A}$ acting on fields which are associated with a representation of some given gauge group. Spacetime tensor fields are best represented as spacetime scalars. Their indices can be associated with representations of the tangent-space gauge group.

For these operators Gilkey has shown that the function $K(f, t)$ has an asymptotic expansion,

$$K(f, t) \sim t^{-D/2} \sum_{N=0}^{\infty} B_{N/2}(f) t^{N/2} \qquad (A.4)$$

in D dimensions. The usefulness of this result follows from the fact that it can be shown that the B coefficients depend only on the function f, the geometry

Table A.1   Heat kernel coefficients b_N with $f = 1$.

Term	Formula
b_0	$\mathrm{tr}(1)$
b_1	$\mathrm{tr}\left(\frac{1}{6}R - X\right)$
b_2	$\mathrm{tr}\left(\frac{1}{72}R^2 - \frac{1}{180}R^{ab}R_{ab} + \frac{1}{180}R^{abcd}R_{abcd} + \frac{1}{30}R_{;a}{}^{a}\right.$
	$\left. -\frac{1}{6}RX + \frac{1}{2}X^2 - \frac{1}{6}X_{;a}{}^{a} + \frac{1}{12}F^{ab}F_{ab}\right)$

and the operator, i.e.

$$B_N = (4\pi)^{-D/2}\int_M b_N(f) + (4\pi)^{-D/2}\int_{\partial M} c_N(f). \qquad (A.5)$$

Explicit forms of some of the b_N and c_N are tabulated in tables A.1 and A.2, restricted to $f = 1$. (Derivative terms involving f can be recovered from conformal rescalings of the metric described later in this appendix.)

The surface terms depend upon the choice of boundary conditions. A general class includes mixtures of Dirichlet and Robin boundary conditions,

$$P_-\phi = 0 \qquad (\psi + \mathbf{n}\cdot\boldsymbol{\nabla})\,P_+\phi = 0 \qquad (A.6)$$

where $P_\pm$ are projection operators and $\mathbf{n}$ is the normal vector. For example,

$$P_-^D = \tfrac{1}{2}(1 + \gamma_5\boldsymbol{\gamma}\cdot\mathbf{n}) \qquad (A.7)$$

for Dirac fermions fixes exactly half of the field components. In table A.2 $P_{+|a}$ denotes the surface derivative of P_+.

The results can be expressed in terms of polynomials in the curvature tensor of the manifold R_{abcd} and the extrinsic curvature of the boundary k_{ab}. Two of these polynomials have a topological origin,

$$q_2 = k^2 - k_{ab}k^{ab} + R - 2R_{ab}n^a n^b \qquad (A.8)$$
$$q_3 = \tfrac{8}{3}k^3 + \tfrac{16}{3}k_a{}^b k_b{}^c k_c{}^a - 8kk_{ab}k^{ab} + 4kR - 8R_{ab}(kn^a n^b + k^{ab})$$
$$+ 8R_{abcd}k^{ac}n^b n^d. \qquad (A.9)$$

The polynomial q_2 is equal to the intrinsic curvature of the boundary by Gauss's equation. According to the Gauss–Bonnet theorem, integrating q_2 over a compact manifold in two dimensions gives the Euler number. In three dimensions the Euler number is one-half the Euler number of the boundary,

$$\chi = \frac{1}{8\pi}\int_{\partial M} d\mu\, q_2. \qquad (A.10)$$

Table A.2 Heat kernel coefficients c_N with $f = 1$.

Term	Formula
$c_{1/2}$	$\frac{1}{2}\pi^{1/2}\text{tr}\,(P_+ - P_-)$
c_1	$\frac{1}{3}\text{tr}\,(k - 6P_+\psi)$
$c_{3/2}^D$	$-\frac{1}{48}q_2 + \frac{1}{32}g_2 + \frac{1}{2}(X - \frac{1}{8}R)$
$c_{3/2}^R$	$\frac{1}{48}q_2 + \frac{1}{32}g_2 - \frac{1}{2}(X - \frac{1}{8}R) + (\psi - \frac{1}{4}k)^2$
$c_{3/2}$	$\pi^{1/2}\text{tr}\left(P_- c_{3/2}^D + P_+ c_{3/2}^R + 48P_{+|a}P_+{}^{|a}\right)$
c_2^D	$-\frac{1}{360}q_3 + \frac{2}{35}g_3 - \frac{1}{3}(X - \frac{1}{6}R)k - \frac{1}{2}\mathbf{n}\cdot\nabla(X - \frac{1}{6}R) + \frac{1}{15}C_{abcd}k^{ac}n^b n^d$
c_2^R	$-\frac{1}{360}q_3 + \frac{2}{45}g_3 - \frac{1}{3}(X - \frac{1}{6}R)k + \frac{1}{2}\mathbf{n}\cdot\nabla(X - \frac{1}{6}R) - \frac{4}{3}(\psi - \frac{1}{3}k)^3$
	$+2(X - \frac{1}{6}R)\psi + (\psi - \frac{1}{3}k)(\frac{2}{45}k^2 - \frac{2}{15}k_{ab}k^{ab}) + \frac{1}{15}C_{abcd}k^{ac}n^b n^d$
c_2	$\text{tr}(P_+ c_2^R + P_- c_2^D - \frac{2}{15}P_{+|a}P_+{}^{|a}\,k - \frac{4}{15}P_{+|a}P_{+|b}k^{ab}$
	$+\frac{4}{3}P_{+|a}P_+{}^{|a}P_+\psi - \frac{2}{3}P_+ P_+{}^{|a}n^b F_{ab})$

In four dimensions, q_3 enters the generalised Gauss–Bonnet theorem for the Euler number,

$$\chi = \frac{1}{32\pi^2}\int_{\mathcal{M}} d\mu\,\lambda_2 + \frac{1}{32\pi^2}\int_{\partial\mathcal{M}} d\mu\,q_3 \qquad (A.11)$$

where $\lambda_2 = R^{abcd}R_{abcd} - 4R^{ab}R_{ab} + R^2$.

Other polynomials are conformally invariant combinations of the extrinsic curvature,

$$g_2 = k_a{}^b k_b{}^a - \frac{1}{2}k^2 \qquad (A.12)$$

$$g_3 = k_a{}^b k_b{}^c k_c{}^a - kk_{ab}k^{ab} + \frac{2}{9}k^3. \qquad (A.13)$$

Conformal variations are defined in the next section.

Combinations of the volume terms for massless fields of various spin, including ghost contributions, are tabulated in table A. Some simplification of these results has been done by assuming that the background spacetime satisfies $R_{ab} = \Lambda g_{ab}$. Results are also given for the four-dimensional ball. For spins less than 2, mixed boundary conditions have been used. For spin 2, the boundary conditions are given in chapter 9.

Table A.3 Heat kernel coefficients $b_2 = \alpha_0 \Lambda^2 + \alpha_1 R_{abcd} R^{abcd}$, B_2 on the sphere $\mathcal{S}$ and ball $\mathcal{B}$. For vector fields $P_-^A = n_a n^b$. For Dirac fields see this appendix and for spin 2, see section 9.5.

Particle	α_0	α_1	$B_2(\mathcal{S})$	$B_2(\mathcal{B})$	P_-	ψ
scalar	$\frac{1}{5}$	$\frac{1}{180}$	$\frac{29}{90}$	$-\frac{1}{180}$	1	0
Dirac	$\frac{2}{15}$	$-\frac{7}{360}$	$\frac{11}{180}$	$\frac{11}{180}$	P_-^D	$\frac{1}{2}k$
photon	$-\frac{4}{45}$	$-\frac{13}{180}$	$-\frac{31}{45}$	$-\frac{31}{90}$	P_-^A	$k_a{}^b$
graviton	$-\frac{2088}{180}$	$\frac{212}{180}$	$-\frac{571}{45}$	$-\frac{241}{90}$		

A.1 Conformal rescaling

A conformal transformation of the metric is a transformation $\widehat{g} = \Omega^2 g$ with a scalar function $\Omega(x)$. The heat kernel expansion has useful transformation properties under conformal transformations. These properties have been used in the past to help determine the expansion coefficients that were listed in the previous section. They are also important for evaluating the quantum averaged trace of the stress–energy tensor of conformally covariant operators, and this is the main concern of this section.

For a conformally covariant operator, $\Delta[\Omega^2 g_{ab}] = \Omega^{c-2} \Delta[g_{ab}] \Omega^{-c}$. The conformal variations of some important quantities are given in table A.1. These can be combined for the conformally covariant scalar operator,

$$-\nabla^2 + \frac{1}{4}\frac{D-2}{D-1}R. \tag{A.14}$$

In four dimensions, the gauge vector operator is invariant and the massless fermion operator is covariant under conformal transformations.

The quantum averaged trace of the stress–energy tensor of a conformally covariant operator can be related to the heat kernel expansion coefficients. The trace is given by variation of the effective action with respect to the metric,

$$\langle T^a{}_a \rangle = -2 \int dx \, g_{ab} \frac{\delta \Gamma}{\delta g_{ab}}. \tag{A.15}$$

If we define

$$\delta_f \Delta = \int d\mu(x) f(x) \Omega(x) \frac{\delta \Delta}{\delta \Omega(x)} \tag{A.16}$$

then

$$\delta_f \Delta = -2f\Delta + c[f, \Delta] \tag{A.17}$$

Table A.4 Conformal variations.

Quantity	With g replaced by $\Omega^2 g$
$\Gamma^a{}_{bc}$	$\Gamma^a{}_{bc} + \Omega^{-1}(\delta^a{}_b\Omega_{;c} + \delta^a{}_c\Omega_{;b} - g_{bc}g^{ad}\Omega_{;d})$
R	$\Omega^{-2}R - 2(D-1)\Omega^{-3}\Omega_{;a}{}^a - (D-1)(D-4)\Omega^{-4}\Omega_{;a}\Omega^{;a}$
∇^2	$\Omega^{-D/2-1}\nabla^2\Omega^{D/2-1} - \frac{1}{2}(D-2)\Omega^{-3}\Omega_{;a}{}^a$
	$-\frac{1}{4}(D-2)(D-4)\Omega^{-4}\Omega_{;a}\Omega^{;a}$

$$\delta_f \mathrm{tr}(e^{-\Delta t}) \;=\; 2\,\mathrm{tr}(ft\Delta e^{-\Delta t}). \tag{A.18}$$

Consider

$$\int d\mu(x)f(x)\langle T^a{}_a(x)\rangle = 2\int d\mu(x)f(x)g_{ab}(x)\frac{\delta\Gamma}{\delta g_{ab}(x)} = \delta_f\Gamma. \tag{A.19}$$

Inserting the regulated expression for the effective action gives,

$$\int d\mu(x)f(x)\langle T^a{}_a(x)\rangle \tag{A.20}$$

$$= -\frac{1}{2}\delta_f\frac{d}{ds}\left\{\frac{1}{\Gamma(s)}\int_0^\infty dt\; t^{s-1}\mu^s\mathrm{tr}\left(e^{-\Delta t}\right)\right\}_{s=0} \tag{A.21}$$

$$= \frac{d}{ds}\left\{\frac{s}{\Gamma(s)}\int_0^\infty dt\; t^{s-1}\mu^s\mathrm{tr}\left(fe^{-\Delta t}\right)\right\}_{s=0} \tag{A.22}$$

$$= \left\{\frac{1}{\Gamma(s)}\int_0^\infty dt\; t^{s-1}\mu^s\mathrm{tr}\left(fe^{-\Delta t}\right)\right\}_{s=0} \tag{A.23}$$

$$= B_{D/2}(f). \tag{A.24}$$

Therefore the conformal anomaly is determined by $B_{D/2}$. (For the conformal scalar wave operator, $X = \frac{1}{6}R$.)

Appendix B

Expansion of the gauge-fixed Einstein action

The gravitational Lagrangian is proportional to the Ricci scalar R. For the quantum theory it is also desirable to add gauge fixing and ghost terms,

$$\mathcal{L} = \mathcal{L}_g + \mathcal{L}_{gf} + \mathcal{L}_{gh}. \tag{B.1}$$

The individual Lagrangians were given in chapter 4,

$$\mathcal{L}_g = \frac{1}{2\kappa^2} R \quad \mathcal{L}_{gf} = -\alpha \widehat{g}^{ab} \mathcal{F}_a \mathcal{F}_b \quad \mathcal{L}_{gh} = 2\overline{c}^a \, s \, \mathcal{F}_a. \tag{B.2}$$

In this appendix we will look at the series expansion of the Lagrangian when the metric $\widehat{\mathbf{g}}$ is close to a classical background metric $\mathbf{g}$, with $\widehat{\mathbf{g}} = \mathbf{g} + \boldsymbol{\gamma}$. Terms up to second order in $\boldsymbol{\gamma}$ will give the quantum theory up to order $\hbar$.

B.1   Curvature expansion

The difference between the connections of the perturbed metric and the background metric is a useful quantity to have available,

$$S^a{}_{bc} = \Gamma^a{}_{bc}(\widehat{\mathbf{g}}) - \Gamma^a{}_{bc}(\mathbf{g}). \tag{B.3}$$

This is a tensor, as is the difference between any two connections. The calculation can be simplified considerably by choosing a coordinate system in which the Christoffel symbol $\Gamma^a{}_{bc}$ for the background metric vanishes,

$$S^a{}_{bc} = \tfrac{1}{2} \widehat{g}^{ad} \left(\gamma_{bd,c} + \gamma_{cd,b} - \gamma_{bc,d} \right). \tag{B.4}$$

In a general frame, the ordinary derivatives can be replaced by the background covariant derivative,

$$S^a{}_{bc} = \tfrac{1}{2} \widehat{g}^{ad} \left(\gamma_{bd;c} + \gamma_{cd;b} - \gamma_{bc;d} \right). \tag{B.5}$$

Table B.1   Linear terms of the Ricci scalar. A semi-colon denotes derivatives with the background metric.

Term	Formula
h_1	$\widehat{g}^{ac}\widehat{g}^{bd}\gamma_{ab;cd}$
h_2	$\widehat{g}^{ab}\widehat{g}^{cd}\gamma_{ab;cd}$

Table B.2   Quadratic terms of the Ricci scalar.

Term	Formula
q_1	$\widehat{g}^{ad}\widehat{g}^{bc}\widehat{g}^{ef}\gamma_{ab;c}\gamma_{de;f}$
q_2	$\widehat{g}^{af}\widehat{g}^{be}\widehat{g}^{cd}\gamma_{ab;c}\gamma_{de;f}$
q_3	$\widehat{g}^{ab}\widehat{g}^{cf}\widehat{g}^{de}\gamma_{ab;c}\gamma_{de;f}$
q_4	$\widehat{g}^{ad}\widehat{g}^{be}\widehat{g}^{cf}\gamma_{ab;c}\gamma_{de;f}$
q_5	$\widehat{g}^{ab}\widehat{g}^{cd}\widehat{g}^{ef}\gamma_{ab;c}\gamma_{de;f}$

Choosing the same frame for the Riemann tensor also gives a tensor equation,

$$R^a{}_{bcd}(\widehat{\mathbf{g}}) = R^a{}_{bcd}(\mathbf{g}) + S^a{}_{bd;c} - S^a{}_{bc;d} + S^a{}_{ec}S^e{}_{bd} - S^a{}_{ed}S^e{}_{bc}. \tag{B.6}$$

The Ricci tensor is obtained by contraction of the Riemann tensor. Derivatives of the inverse metric can be simplified by

$$\widehat{g}^{ab}{}_{;c} = -\widehat{g}^{ad}\widehat{g}^{eb}\gamma_{de;c}. \tag{B.7}$$

After substitutions,

$$R_{ab}(\widehat{\mathbf{g}}) = R_{ab}(\mathbf{g}) + \widehat{g}^{cd}H_{abcd} + \widehat{g}^{cd}\widehat{g}^{ef}Q_{abcdef} \tag{B.8}$$

where

$$H_{abcd} = \tfrac{1}{2}(\gamma_{ca;bd} + \gamma_{cb;ad} - \gamma_{ab;cd} - \gamma_{cd;ab}) \tag{B.9}$$

and

$$Q_{abcdef} = \tfrac{1}{2}(\gamma_{cd;e} - 2\gamma_{ce;d})(\gamma_{f(a;b)} - \tfrac{1}{2}\gamma_{ab;f}) + \tfrac{1}{4}\gamma_{de;b}\gamma_{cf;a} + \gamma_{b[d;e]}\gamma_{a[c;f]}. \tag{B.10}$$

The Ricci scalar can now be written

$$R(\widehat{\mathbf{g}}) = \widehat{g}^{ab}R_{ab}(\mathbf{g}) + h_1 - h_2 - q_1 - \tfrac{1}{2}q_2 - \tfrac{1}{4}q_3 + \tfrac{3}{4}q_4 + q_5 \tag{B.11}$$

with linear terms and quadratic terms given in tables B.1 and B.2.

Integrals of the Ricci scalar can be simplified a little using

$$h_1 = q_1 + q_2 - \tfrac{1}{2}q_5 + j_{1\ ;a}^{a} \qquad j_1^{d} = \widehat{g}^{ac}\widehat{g}^{bd}\gamma_{ab;c} \qquad (\text{B.12})$$

$$h_2 = q_4 + q_5 - \tfrac{1}{2}q_3 + j_{2\ ;a}^{a} \qquad j_2^{d} = \widehat{g}^{ab}\widehat{g}^{cd}\gamma_{ab;c}. \qquad (\text{B.13})$$

The Ricci scalar is therefore

$$R(\widehat{g}) = \widehat{g}^{ab} R_{ab}(g) - \tfrac{1}{2}q_5 - \tfrac{1}{4}q_4 + \tfrac{1}{2}q_2 + \tfrac{1}{4}q_3 + (j_1^{a} - j_2^{a})_{;a}. \qquad (\text{B.14})$$

The q_3 and q_4 terms are in a standard form for Lagrangians, but the q_2 and q_5 terms are not.

B.2   Boundary terms

The total derivative terms in the Ricci tensor can be dropped from the Lagrangian if the manifold has no boundary. If the manifold has a boundary then these terms can be removed by a combination of imposing boundary conditions and modifying the original action.

First some notation. The normal to the boundary $\partial\mathcal{M}$ will be denoted by $\mathbf{n}$. The metric $\mathbf{g}$ induces a metric

$$\mathbf{h} = \mathbf{g} - \mathbf{n} \otimes \mathbf{n} \qquad (\text{B.15})$$

on $\partial\mathcal{M}$. This metric projects onto the space orthogonal to the normal, $g^{ab}n_a h_{bc} = 0$. Other important properties of $\partial\mathcal{M}$ are the extrinsic curvature,

$$k_{ab} = h_a{}^{c} h_b{}^{d} n_{c;d} \qquad (\text{B.16})$$

and the surface derivative $|$.

The terms that arise from the Ricci tensor can be partially rewritten in terms of surface quantities,

$$n^{a}(j_{1a} - j_{2a}) = n^{a}(\widehat{j}_{1a} - \widehat{j}_{2a}) \qquad (\text{B.17})$$

where

$$\widehat{j}_{1a} = \widehat{h}^{bc}\gamma_{ab;c} \qquad \widehat{j}_{2a} = \widehat{h}^{bc}\gamma_{bc;a}. \qquad (\text{B.18})$$

The first of these terms depends only on the surface components of γ,

$$n^{a}\widehat{j}_{1a} = \widehat{k}^{ab}\gamma_{ab} - (n^{b}\gamma_{ab})^{|a}. \qquad (\text{B.19})$$

The second term $\widehat{j}_2$ depends upon normal derivatives of γ.

If the metric is held fixed on the boundary, corresponding to Dirichlet boundary conditions on the metric variation, then $\widehat{j}_2$ needs to be eliminated. The trace of the extrinsic curvature does the trick, for equation (B.3) implies

$$k(\widehat{g}) - k(g) = h^{a}{}_{c}S^{c}{}_{ab}n^{b} = \tfrac{1}{2}n^{a}\widehat{j}_{2a}. \qquad (\text{B.20})$$

Therefore the original gravitational action should be modified to

$$I = -\frac{1}{2\kappa^2} \int_{\mathcal{M}} d\mu \, R - \frac{1}{\kappa^2} \int_{\partial\mathcal{M}} d\mu \, k. \tag{B.21}$$

After the expansion to linear order, the surface term becomes

$$-\int_{\partial\mathcal{M}} d\mu \, (\widehat{k}^{ab} - \widehat{h}^{ab} k)\gamma_{ab} \tag{B.22}$$

which vanishes for these boundary conditions.

B.3   Ghosts

Now we are ready for the gauge fixing and ghost terms. The specific form of the gauge fixing functional should be chosen with the aim of keeping the action simple. The gauge fixing functional

$$\mathcal{F}_a = \tfrac{1}{2}\kappa^{-1}\widehat{g}^{bc} \left(\gamma_{ab;c} - \tfrac{1}{2}\gamma_{bc;a}\right) \tag{B.23}$$

has this useful property. The contribution to the Lagrangian becomes

$$\mathcal{L}_{gf} = \tfrac{1}{4}\kappa^{-2}\alpha \left(-q_1 + q_5 - \tfrac{1}{4}q_3\right). \tag{B.24}$$

For the ghost terms in the action, we need

$$s\,\mathcal{F}_a = \tfrac{1}{2}\kappa^{-1}\widehat{g}^{bc}(c_{a;bc} + c_{b;ac} - c_{b;ca}). \tag{B.25}$$

Consequently,

$$\mathcal{L}_{gh} = \kappa^{-1}\overline{c}^a \, \Delta_{Ga}{}^b c_b, \tag{B.26}$$

where the ghost operator is given later in table B.3.

B.4   Truncation

No approximations have been made so far but extracting terms of a particular order requires expanding the inverse metric and the measure in powers of **h**. In the case of the measure we use

$$\det(\mathbf{g} + \boldsymbol{\gamma}) = \exp\left\{\text{tr}\log(\mathbf{g} + \boldsymbol{\gamma})\right\}. \tag{B.27}$$

This gives

$$\sqrt{\widehat{g}} = \sqrt{g}\left\{1 - \tfrac{1}{2}h + g^{(ab)(cd)}\gamma_{ab}\gamma_{cd} + \ldots\right\} \tag{B.28}$$

where g^{ab} are the components of $\mathbf{g}^{-1}$ and,

$$g^{(ab)(cd)} = \tfrac{1}{2}\left(g^{ac}g^{bd} + g^{ad}g^{bc} - g^{ab}g^{cd}\right). \qquad (B.29)$$

This is an important tensor in the theory of metric perturbations. It can be regarded as a metric with indices (ab) and (cd). The inverse metric is

$$g_{(ab)(cd)} = \tfrac{1}{2}\left(g_{ac}g_{bd} + g_{ad}g_{bc} - g_{ab}g_{cd}\right) \qquad (B.30)$$

and the trace

$$g_{(ab)(cd)}g^{(ab)(cd)} = 10 \qquad (B.31)$$

gives the number of independent components of the original metric. Furthermore the Einstein tensor is given by $G^{ab} = g^{(ab)(cd)}R_{cd}$.

The quadratic terms in the action can be put into operator form using integration by parts,

$$q_3 = -g^{ab}g^{cd}\gamma_{ab}\nabla^2\gamma_{cd} + j_3{}^a{}_{;a} + \mathcal{L}_i^{(3)} \qquad (B.32)$$

$$q_4 = -g^{ac}g^{bd}\gamma_{ab}\nabla^2\gamma_{cd} + j_4{}^a{}_{;a} + \mathcal{L}_i^{(4)} \qquad (B.33)$$

$$(q_1 - q_2) = -g^{ab}g^{cd}g^{ef}\gamma_{ae}(\gamma_{cf;db} - \gamma_{cf;bd}) + j_0{}^a{}_{;a}$$

$$+ (\mathcal{L}_i^{(1)} - \mathcal{L}_i^{(2)}). \qquad (B.34)$$

where the currents j_n vanish on the boundary and $\mathcal{L}_i$ are higher-order interaction terms.

Table B.3   Fluctuation operators for the graviton and gravity ghost.

Operator	Result
$\Delta_L^{(ab)(cd)}$	$g^{(ab)(cd)}(-\nabla^2 + R) - 2g^{(ab)(ef)}R^c{}_{egf}g^{dg}$
	$-g^{(ab)(ed)}R_e{}^c - g^{(ab)(ce)}R_e{}^d$
$\Delta_\alpha^{(ab)(cd)}$	$\tfrac{1}{2}(1 - \alpha)g^{(ab)(eg)}g^{(cd)(fh)}g_{gh}\nabla_e\nabla_f$
$\Delta_{Ga}{}^b$	$-\delta_a{}^b\nabla^2 + R_a{}^b$

Gathering the terms (B.14) and (B.28) together gives the total Lagrangian

$$\mathcal{L} = -\tfrac{1}{2}\kappa^{-2}R + \tfrac{1}{2}\kappa^{-2}G^{ab}\gamma_{ab} + \tfrac{1}{8}\kappa^{-2}\gamma_{ab}\Delta_g^{(ab)(cd)}\gamma_{cd} + \mathcal{L}_i + \mathcal{L}_{gh}. \qquad (B.35)$$

The graviton operator can be divided into two components Δ_L and Δ_α, with the explicit expressions given in table B.3. The first term is the fluctuation operator in the gauge $\mathcal{F}_a = 0$. This operator, called the Lichnerovitz operator, can be used to describe the motion of gravitons on a curved background spacetime. The second term contains the dependence upon the gauge fixing parameter, α.

B.5   Sources

In many situations it will be necessary to include extra fields along with gravity. The expansion of the Riemann tensor is unaffected, but there are likely to be extra terms involving the metric variation from the matter field Lagrangian. In the case of a massless scalar field and a cosmological constant,

$$\mathcal{L}_m = \tfrac{1}{2} \boldsymbol{\nabla}\phi \cdot \boldsymbol{\nabla}\phi \quad \mathcal{L}_\Lambda = -\kappa^{-2}\Lambda. \tag{B.36}$$

If the scalar field has its own background, $\boldsymbol{\nabla}\phi = \mathbf{E} + \boldsymbol{\nabla}\xi$, where ξ represents the fluctuation. Under a BRST transformation, $s\xi = -c^a E_a$.

Both the metric and scalar fluctuations can be combined into a single fluctuation vector,

$$\boldsymbol{\eta} = ((1/2\kappa)\gamma_{ab}, \xi). \tag{B.37}$$

Some simplification can be achieved by altering the gauge fixing term,

$$\mathcal{F}_a = \tfrac{1}{2}\kappa^{-1}\widehat{g}^{bc}\left(\gamma_{ab;c} - \tfrac{1}{2}\gamma_{bc;a}\right) - \kappa E_a \xi. \tag{B.38}$$

The second-order gauge-fixed Lagrangian $\mathcal{L}_2$ for the graviton and scalar,

$$\mathcal{L}_2 = \tfrac{1}{2}\boldsymbol{\eta} \cdot \boldsymbol{\Delta}\boldsymbol{\eta}. \tag{B.39}$$

The operator is now

$$\boldsymbol{\Delta} = \begin{pmatrix} \Delta_{L(ab)}^{(cd)} - 2\Lambda\delta_{(ab)}^{(cd)} & -2\kappa\nabla_a E_b \\ -2\kappa\nabla^c E^d & -\nabla^2 - 2\kappa^2 \mathbf{E} \cdot \mathbf{E} \end{pmatrix} \tag{B.40}$$

in Feynman gauge ($\alpha = 1$).

Appendix C

Curvature forms and 3+1 decompositions

The Hamiltonian for gravity will be found in this appendix by using a basis of 1-forms. Alternative discussions of the canonical decomposition can be found in textbooks on relativity. The use of 1-forms provides a way of describing curvature which is efficient and can be generalised to Yang–Mills theory.

Denote an orthonormal basis of 1-forms by $\boldsymbol{\omega}^a$ for $a = 0, \dots, 3$. These forms are linear maps on the tangent space, $\boldsymbol{\omega}^a(\mathbf{X}) = X^a$. The orthonormality means that the metric is $\mathbf{g} = \eta_{ab}\boldsymbol{\omega}^a \otimes \boldsymbol{\omega}^b$.

Higher-order forms are built up using the wedge product

$$\boldsymbol{\omega}^a \wedge \boldsymbol{\omega}^b = \tfrac{1}{2}\left(\boldsymbol{\omega}^a \otimes \boldsymbol{\omega}^b - \boldsymbol{\omega}^b \otimes \boldsymbol{\omega}^a\right). \tag{C.1}$$

Linear combinations of products of p 1–forms make up the space of rank-p forms,

$$\boldsymbol{\alpha} = \alpha_{ab\dots c}\boldsymbol{\omega}^a \wedge \boldsymbol{\omega}^b \dots \wedge \boldsymbol{\omega}^c. \tag{C.2}$$

Forms may also be differentiated by the exterior derivative operator $\mathbf{d}$, which satisfies

$$\mathbf{d}(\boldsymbol{\alpha} \wedge \boldsymbol{\beta}) = \mathbf{d}\boldsymbol{\alpha} \wedge \boldsymbol{\beta} + (-1)^p \boldsymbol{\alpha} \wedge \mathbf{d}\boldsymbol{\beta} \tag{C.3}$$

and $\mathbf{d}f = f_{,\mu}\mathbf{dx}^\mu$, comma denoting partial differentiation, for scalars.

In the general theory of relativity curvature and torsion tensors are introduced to represent the physical effects of geometry. These tensors are defined concisely in terms of the connection $\boldsymbol{\nabla}$ acting on tangent vectors

$$\mathbf{T}(\mathbf{X}, \mathbf{Y}) = \boldsymbol{\nabla}_X \mathbf{Y} - \boldsymbol{\nabla}_Y \mathbf{X} - [\mathbf{X}, \mathbf{Y}], \tag{C.4}$$

$$\mathbf{R}(\mathbf{X}, \mathbf{Y})\mathbf{Z} = \boldsymbol{\nabla}_X \boldsymbol{\nabla}_Y \mathbf{Z} - \boldsymbol{\nabla}_Y \boldsymbol{\nabla}_X \mathbf{Z} + \boldsymbol{\nabla}_{[X,Y]}\mathbf{Z}. \tag{C.5}$$

The torsion tensor is set to zero in the general theory.

Both of these definitions are antisymmetric in $\mathbf{X}$ and $\mathbf{Y}$. This means that it is possible to define torsion and curvature 2-forms,

$$\mathbf{T}^a = \tfrac{1}{2}T^a{}_{bc}\boldsymbol{\omega}^b \wedge \boldsymbol{\omega}^c \tag{C.6}$$

$$\mathbf{R}^a{}_b = \tfrac{1}{2} R^a{}_{bcd} \boldsymbol{\omega}^c \wedge \boldsymbol{\omega}^d \tag{C.7}$$

using the basis of forms.

Denote the connection in the 1-form basis by

$$\boldsymbol{\omega}^a{}_b = \Gamma^a{}_{bc}\boldsymbol{\omega}^c = \boldsymbol{\omega}^a \left(\nabla_c \boldsymbol{\omega}_b \right) \boldsymbol{\omega}^c \tag{C.8}$$

where $\boldsymbol{\omega}_a$ is the dual basis of vectors. It then follows that

$$\mathbf{T}^a = \mathbf{d}\boldsymbol{\omega}^a + \boldsymbol{\omega}^a{}_b \wedge \boldsymbol{\omega}^b \tag{C.9}$$

$$\mathbf{R}^a{}_b = \mathbf{d}\boldsymbol{\omega}^a{}_b + \boldsymbol{\omega}^a{}_c \wedge \boldsymbol{\omega}^c{}_b \tag{C.10}$$

Deriving these formulae is a beneficial exercise.

The relationship between the metric and the connection is the condition $\nabla \mathbf{g} = 0$, which implies that

$$\eta_{ac}\boldsymbol{\omega}^c{}_b + \eta_{bc}\boldsymbol{\omega}^c{}_a = 0. \tag{C.11}$$

Furthermore, the torsion tensor is taken to vanish. This allows the connection forms to be written explicitly in terms of commutators $[\boldsymbol{\omega}_a, \boldsymbol{\omega}_b] = f_{ab}{}^c \boldsymbol{\omega}_c$,

$$\omega_{ab} = \tfrac{1}{2}(f_{abc} - f_{bca} - f_{cab})\boldsymbol{\omega}^c. \tag{C.12}$$

The indices in this equation have been lowered with η_{ab}. Alternatively,

$$\boldsymbol{\omega}^a = \omega^a{}_\mu \mathbf{dx}^\mu \quad \Longrightarrow \quad \boldsymbol{\omega}^a{}_b = -\omega^a{}_{\mu,\nu}\omega_b{}^\nu \mathbf{dx}^\mu \tag{C.13}$$

using a coordinate basis.

C.1 $3+1$ decomposition

Now consider a spacetime which is sliced into spatial hypersurfaces Σ_t of constant time coordinate t. The normal form to these slices will be taken as one member of the orthonormal basis of 1-forms, and this is proportional to $\mathbf{dt}$. The other members of the basis will be linear combinations of $\mathbf{dt}$ and the remaining coordinate forms $\mathbf{dx}^i$,

$$\boldsymbol{\omega}^0 = N\mathbf{dt} \tag{C.14}$$

$$\boldsymbol{\omega}^m = \omega^m{}_i (N^i \mathbf{dt} + \mathbf{dx}^i). \tag{C.15}$$

The functions N and N^i defined here are called the lapse and shift functions. They depend on the choice of coordinates. Dual vectors are found to be

$$\boldsymbol{\omega}_0 = \frac{1}{N}\frac{\partial}{\partial t} - \frac{N^i}{N}\frac{\partial}{\partial x^i} \tag{C.16}$$

$$\boldsymbol{\omega}_m = \omega^i{}_m \frac{\partial}{\partial x^i}. \tag{C.17}$$

If the metric is written in the coordinate basis, then

$$\mathbf{g} = \eta_{ab}\boldsymbol{\omega}^a \otimes \boldsymbol{\omega}^b = \begin{pmatrix} -N^2 + N_k N^k & N_j \\ N_i & h_{ij} \end{pmatrix} \tag{C.18}$$

$$\mathbf{g}^{-1} = \eta^{ab}\boldsymbol{\omega}_a \otimes \boldsymbol{\omega}_b = \begin{pmatrix} -N^{-2} & N^{-2}N^j \\ N^{-2}N^i & h^{ij} - N^{-2}N^iN^j \end{pmatrix} \tag{C.19}$$

where $h_{ij} = \eta_{mn}\omega^m{}_i\omega^n{}_j$.

The next step is to decompose the connection forms. Consider

$$\omega^0{}_m = k_{mn}\boldsymbol{\omega}^n + a_m\boldsymbol{\omega}^0 \tag{C.20}$$

$$\omega^m{}_n = \sigma^m{}_n - b^m{}_n\boldsymbol{\omega}^0. \tag{C.21}$$

The spatial tensor k_{mn} is called the extrinsic curvature of the spatial surfaces. The $\sigma^m{}_n$ are connection forms for the metric η_{mn} on the spatial hypersurfaces, and b_{mn} is an antisymmetric tensor.

Now, take a look at

$$d\boldsymbol{\omega}^0 = -k_{mn}\boldsymbol{\omega}^n \wedge \boldsymbol{\omega}^m - a_{mn}\boldsymbol{\omega}^m \wedge \boldsymbol{\omega}^0 \tag{C.22}$$

$$d\boldsymbol{\omega}^m = b^m{}_n\boldsymbol{\omega}^n \wedge \boldsymbol{\omega}^0 - k^m{}_n\boldsymbol{\omega}^n \wedge \boldsymbol{\omega}^0 - \sigma^m{}_n \wedge \boldsymbol{\omega}^n. \tag{C.23}$$

With the particular form of $\boldsymbol{\omega}^0$ in (C.14), which is hypersurface orthogonal, it follows that we have

$$\omega^m{}_i a_m = \frac{N_{|i}}{N}. \tag{C.24}$$

A vertical bar denotes covariant derivatives using the three-dimensional connection. It also follows that k_{mn} is symmetric in the indices m and n.

Inserting the coordinate forms for $\boldsymbol{\omega}^m$ into the next equation gives

$$k_{mn} = N^{-1}\left(\dot{\omega}_{(mi}\omega_{n)}{}^i - N_{(m|n)}\right) \tag{C.25}$$

$$b_{mn} = N^{-1}\left(\dot{\omega}_{[mi}\omega_{n]}{}^i - N_{[m|n]}\right). \tag{C.26}$$

It is often convenient to introduce a coordinate basis metric $h_{ij} = \omega^m{}_i\omega_{nj}$, in which case the intrinsic curvature can be rewritten as

$$\omega^m{}_i\omega^n{}_j k_{mn} = \frac{1}{2N}\left(\frac{\partial h_{ij}}{\partial t} - N_{i|j} - N_{j|i}\right). \tag{C.27}$$

The curvature forms can be expressed as

$$\mathbf{R}^m{}_0 = d\boldsymbol{\omega}^m{}_0 + \boldsymbol{\omega}^m{}_n \wedge \boldsymbol{\omega}^n{}_0 \tag{C.28}$$

$$\mathbf{R}^m{}_n = d\boldsymbol{\omega}^m{}_n + \boldsymbol{\omega}^m{}_0 \wedge \boldsymbol{\omega}^0{}_n + \boldsymbol{\omega}^m{}_p \wedge \boldsymbol{\omega}^p{}_n \tag{C.29}$$

After substituting in the connection forms it is possible to read off components of the Riemann tensor,

$$R^m{}_{0n0} = k^m{}_p k^p{}_n + a^m{}_{|n} + a^m a_n - 2k^m{}_p b^p{}_n - \omega_0(k^m{}_n) \quad (C.30)$$

$$R^m{}_{0np} = k^m{}_{n|p} - k^m{}_{p|n} \quad (C.31)$$

$$R^m{}_{npq} = r^m{}_{npq} + k^m{}_p k_{nq} - k^m{}_q k_{np} \quad (C.32)$$

where $r^m{}_{npq}$ is the curvature tensor of the hypersurfaces. The first equation is a form of the Raychaudhuri equation for the focusing of the vector field normal to the hypersurfaces. The second equation is called the Codazzi equation and the third is called Gauss's equation.

C.2   Gravitational action

The Einstein action is given by

$$S = \frac{1}{2\kappa^2} \int_{\mathcal{M}} d\mu\, R + \frac{1}{\kappa^2} \int_{\partial\mathcal{M}} d\mu\, k, \quad (C.33)$$

where $\kappa^2 = 8\pi G$. A boundary term of this form has to be included if there is a boundary and the metric is fixed on the boundary.

The volume term can be written in terms of the Riemann tensor,

$$\sqrt{-g}\, R = N\sqrt{h}\left(2R^{0m}{}_{0m} + R^{mn}{}_{mn}\right). \quad (C.34)$$

Using Gauss's equation yields

$$R^{mn}{}_{mn} = r + k^2 - k^{ij} k_{ij}. \quad (C.35)$$

Coordinate expansions of a_i and ω_0 can be inserted into equation (C.30) for $R^m{}_{0m0}$ to get

$$R^m{}_{0m0} = k^m{}_p k^p{}_m - k^2 + N^{-1}\left(N^{|i}{}_i - h^{-1/2}\partial(k\sqrt{h})/\partial t - (N^i k)_{|i}\right). \quad (C.36)$$

If the manifold has no boundary, the terms in parentheses can be integrated out of the action using the divergence theorem, leaving

$$S = \frac{1}{2\kappa^2} \int \left(r - k^2 + k^{ij} k_{ij}\right) N\, d\mu(x) dt, \quad (C.37)$$

which is to be used in conjunction with the coordinate expression for k_{ij}.

If the manifold is bounded by two of the space-like slices then the derivative terms are cancelled by the boundary term in the original action. If there are spatial boundaries, then there is a residual integral over $\partial\Sigma \cap \Sigma$ of the extrinsic curvature $\hat{k}$ of $\partial\Sigma$ embedded in Σ.

Table C.1   Hamiltonian density for relativistic fields.

Field	$\mathcal{H}$	$\mathcal{H}_i$
metric	$2\kappa^2 \, h^{-1/2} \left(p^{ij} p_{ij} - \frac{1}{2} p^2 \right) - r \, h^{1/2}/(2\kappa^2)$	$-2h_{ik} p^{kj}{}_{|j}$
vector	$\frac{1}{2} h^{-1/2} \left(\boldsymbol{E} \cdot \boldsymbol{E} + \boldsymbol{B} \cdot \boldsymbol{B} \right)$	$h^{1/2} \boldsymbol{E} \times \boldsymbol{B}$
scalar	$h^{-1/2} \pi \pi^\dagger + h^{1/2} h^{ij} \phi_{|i} \phi^\dagger_{|j} + h^{1/2} V(\phi)$	$\pi^\dagger \phi_{|i} + \pi \phi^\dagger_{|i}$

C.3   Hamiltonians

The decomposed action allows for a Hamiltonian expression of gravitational dynamics. The momentum density is obtained by differentiating the Lagrangian density inside the integral with respect to $\partial h_{ij}/\partial t$,

$$p^{ij} = \frac{1}{2\kappa^2} \, h^{1/2} \left(k^{ij} - h^{ij} k \right). \tag{C.38}$$

The transformation to p^{ij} then gives

$$S = \int \left(\dot{h}_{ij} p^{ij} - N\mathcal{H} - N^i \mathcal{H}_i \right) d^3x \, dt \tag{C.39}$$

where $\mathcal{H}$ and $\mathcal{H}_i$ are given in table C.3. There are no momenta conjugate to the variables N and N^i. Variation of the action with respect to these variables gives a subset of the Einstein equations, called the constraint equations, $\mathcal{H} = 0$ and $\mathcal{H}_i = 0$.

Matter field actions can be decomposed in a similar way. Results for a scalar field, with momentum π, are given in table C.3. For electromagnetic fields the action can be similarly decomposed,

$$S = \int \left(\dot{A}_i E^i - A_0 E^i{}_{|i} - N\mathcal{H} - N^i \mathcal{H}_i \right) d^3x \, dt. \tag{C.40}$$

The momentum conjugate to the three-vector potential $\boldsymbol{A}$ is the electric field $\boldsymbol{E}$, and the magnetic field

$$\boldsymbol{B} = h^{1/2} \nabla \times \boldsymbol{A} \tag{C.41}$$

in the notation $\boldsymbol{A} \times \boldsymbol{B} = \epsilon^{ijk} A_j B_k$. In this example there is no momentum conjugate to A_0, and variation of the action with respect to A_0 gives the constraint equation which is usually called Gauss's law.

Appendix D

Spherical harmonics and hyperspheres

This appendix contains some useful properties of the unit sphere. In $(D+1)$-dimensional flat space with Cartesian coordinates $\mathbf{x} = (x, y, z, \ldots)$, the unit sphere is the surface

$$\mathbf{x} \cdot \mathbf{x} = 1. \tag{D.1}$$

Angular coordinates will be denoted by $(\phi, \theta, \chi, \ldots)$, with $0 \leq \phi < 2\pi$ and $0 < \theta, \chi, \ldots < \pi$. On the sphere, where $\mathbf{x}$ is denoted by $\widehat{\mathbf{x}}$, the relationship between the two sets of coordinates is

$$
\begin{array}{cc}
\widehat{x} & \cos\phi \sin\theta \sin\chi \ldots \\
\widehat{y} & \sin\phi \sin\theta \sin\chi \ldots \\
\widehat{z} & \cos\theta \sin\chi \ldots \\
\text{etc.}
\end{array}
\tag{D.2}
$$

The Cartesian vector $\widehat{\mathbf{x}}$ can also be identified with the unit normal to the sphere. In spherical polar coordinates $\mathbf{x} = r\widehat{\mathbf{x}}$, and the induced metric on the sphere is simply $d\widehat{\mathbf{x}} \cdot d\widehat{\mathbf{x}}$. The metric for the ordinary sphere of two intrinsic dimensions would be

$$ds^2 = d\theta^2 + \sin^2\theta \, d\phi^2 \tag{D.3}$$

in angular coordinates. Also

$$ds_D^2 = d\chi^2 + \sin^2\chi \, ds_{D-1}^2 \tag{D.4}$$

for the metric in D dimensions.

D.1 The hypersphere

The hypersphere is the surface of the unit sphere in four-dimensional flat space. The hypersphere has special properties because it is also the group manifold of SU(2). An element of SU(2) can be labelled by Euler angles,

$$\mathbf{U} = \begin{pmatrix} \cos(\vartheta/2)\, e^{i(\psi+\phi)/2} & \sin(\vartheta/2)\, e^{-i(\psi-\phi)/2} \\ -\sin(\vartheta/2)\, e^{i(\psi-\phi)/2} & \cos(\vartheta/2)\, e^{-i(\psi+\phi)/2} \end{pmatrix} \tag{D.5}$$

where $0 < \psi, \phi < 2\pi$ and $0 < \vartheta < \pi$. These angles also form coordinates on $\mathbf{S}^3$.

The diffeomorphisms generated by multiplying on the left by a fixed element of SU(2) leave a set of forms invariant, namely $\mathbf{U}^{-1}\mathbf{dU}$. We can define a basis of left-invariant forms $\boldsymbol{\omega}^i$ by expanding this matrix in Pauli matrices $\boldsymbol{\sigma}_i$,

$$\mathbf{U}^{-1}\mathbf{dU} = \tfrac{1}{2} i \boldsymbol{\omega}^i \boldsymbol{\sigma}_i \tag{D.6}$$

where

$$\boldsymbol{\sigma}_1 = \begin{pmatrix} 0 & 1 \\ 1 & 0 \end{pmatrix} \quad \boldsymbol{\sigma}_2 = \begin{pmatrix} 0 & -i \\ i & 0 \end{pmatrix} \quad \boldsymbol{\sigma}_3 = \begin{pmatrix} 1 & 0 \\ 0 & -1 \end{pmatrix}. \tag{D.7}$$

Explicit expressions for the forms in terms of the Euler angles are

$$\boldsymbol{\omega}^1 = -\sin\psi\, \mathbf{d\vartheta} + \cos\psi \sin\vartheta\, \mathbf{d\phi}, \tag{D.8}$$

$$\boldsymbol{\omega}^2 = \cos\psi\, \mathbf{d\vartheta} + \sin\psi \sin\vartheta\, \mathbf{d\phi}, \tag{D.9}$$

$$\boldsymbol{\omega}^3 = \mathbf{d\psi} + \cos\vartheta\, \mathbf{d\phi}. \tag{D.10}$$

These left-invariant forms are normalised to satisfy the Cartan relations $\mathbf{d\omega}^i = \epsilon^i{}_{jk} \boldsymbol{\omega}^i \wedge \boldsymbol{\omega}^j$, which means that with respect to the metric

$$ds^2 = \tfrac{1}{2}\mathrm{tr}(\mathbf{U}^{-1}\mathbf{dU} \otimes \mathbf{U}^{-1}\mathbf{dU}) = \tfrac{1}{4}\sum_i \boldsymbol{\omega}^i \otimes \boldsymbol{\omega}^i \tag{D.11}$$

they are only normalised up to factors of 2.

D.2 Harmonics

Tensor harmonics $\mathbf{Q}$ on the D-dimensional sphere are defined to be tensor functions that satisfy

$$-\nabla^2 \mathbf{Q} = \lambda \mathbf{Q}. \tag{D.12}$$

They can be classified as scalar, vector, tensors of rank 2 etc.

The way in which the harmonics will be labelled here is shown in table D.1. The lowest-rank harmonics are scalar eigenfunctions Q with eigenvalues λ^S. The space of vector eigenfunctions decomposes into gradients of the scalar harmonics Q_i and divergence-free or transverse eigenfunctions P_i. As we increase the rank, a set of trace-free and diverge-free tensors emerges. In table D.1 these are labelled 'TT'.

Table D.1   Tensor harmonics and eigenvalues on the sphere.

Rank	TT	Non–TT	
0	Q		
	λ^S		
1	P_i	$Q_{;i}$	
	λ^P	$\lambda^S - D + 1$	
2	G_{ij}	$P_{i;j} + P_{j;i}$	$Q_{;ij} + (\lambda^S/D)g_{ij}Q$
	λ^G	$\lambda^P - D + 1$	$\lambda^S - 2D$

The scalar harmonics can be constructed from polynomials p_n of order n in the Cartesian coordinates x^a. The $(D + 1)$-dimensional Laplacian acting on scalars decomposes on the sphere into

$$\delta^{ab}\nabla_a\nabla_b = r^{-D}\partial_r r^D \partial_r + r^{-2}\nabla^2 \qquad (D.13)$$

Polynomials annihilated by the Laplacian on the left are called harmonic polynomials. If any harmonic polynomial is written in the form $p_n = r^n Q_n(\widehat{\mathbf{x}})$, then

$$n(n + D - 1)Q_n + \nabla^2 Q_n = 0. \qquad (D.14)$$

The Q_n are therefore spherical harmonics with eigenvalues $\lambda^S = n(n+D-1)$. The degeneracy of these eigenvalues is given by the number of independent harmonic polynomials of degree n.

Higher-rank harmonics can be constructed by combining harmonic polynomials in D dimensions with a set of projection matrices $\omega_i{}^a$, which project from the Cartesian frame onto the tangent space of the sphere $(\omega_i{}^a = \widehat{x}^a{}_{,i})$.

D.3   Harmonics on the two-sphere

Spherical harmonics are associated with representations of the rotation group of three-dimensional space. The most frequently used basis of normalised spherical harmonics is the set $Y_{lm}(\theta, \phi)$, associated with complex irreducible representaions.

Vector and tensor harmonics can also be built up using rotation group theory. Let $\mathbf{L}$ be the (reduced) angular momentum operator and $\mathbf{S}$ a generator

Table D.2 Low order Cartesian TT harmonics on the sphere.

Rank	$n = 0$	$n = 1$	$n = 2$
0	$Q = 1$	$Q^a = \widehat{x}^a$	$Q^{ab} = \widehat{x}^{(a}\widehat{x}^{b)} - \delta^{ab}/D$
1		$P_i^{ab} = \omega_i{}^{[a}\widehat{x}^{b]}$	$P_i^{abc} = \omega_i{}^{[a}Q^{b]c}$
2			$G_{ij}^{abcd} = \omega_{(i}{}^{a}\omega_{j)}{}^{[b}Q^{c]d}$

Table D.3 Eigenvalues for TT spherical harmonics in D dimensions.

Rank	Eigenvalue
0	$n(n + D - 1)$
$\frac{1}{2}$	$(n + \frac{1}{2}D)^2 - \frac{1}{4}D(D - 1)$
1	$n(n + D - 1) - 1$
2	$n(n + D - 1) - 2$

of rotations on the tensor harmonic indices. The matrix $\mathbf{S}$ acts as a connection allowing the covariant derivative to be written in terms of $\mathbf{J} = \mathbf{L} + \mathbf{S}$,

$$i\nabla_i = \omega_i{}^a \mathbf{J}_a. \tag{D.15}$$

The eigenvalues of the Laplacian therefore coincide with eigenvalues of J^2, which are $j(j + 1)$. The harmonics can be obtained by combining spherical harmonics of rank l with irreducible tensors of rank s using Clebsch–Gordan coefficients.

D.4 Harmonics on the hypersphere

The eigenfunctions of the Laplacian on S^3 can also be classified by the $SU(2)$ symmetry. We associate an angular momentum operator $\mathbf{J}$ with the Lie derivative along the left-invariant form $\boldsymbol{\omega}_i$, $J_i = i\mathcal{L}_{\omega_i}$. This we decompose into a part $\mathbf{S}$ which acts on the frame indices of tensors and a part $\mathbf{L}$ which acts on the functional part, $\mathbf{J} = \mathbf{L} + \mathbf{S}$. The covariant derivative (in the left-invariant basis) can also be written in terms of these operators and turns out

Table D.4 Degeneracies for TT spherical harmonics in D dimensions.

Rank	Degeneracy
0	$\dfrac{(2n + D - 1)\Gamma(D + n - 1)}{\Gamma(D)\Gamma(n + 1)}$
$\frac{1}{2}$	$\dfrac{\Gamma(D + l)}{\Gamma(D)\Gamma(n + 1)}$
1	$\dfrac{n(n + D - 1)(2n + D - 1)\Gamma(D + n - 2)}{\Gamma(D - 1)\Gamma(n + 2)}$
2	$\dfrac{(D + 1)(D - 2)(D + n)(n - 1)(2n + D - 1)\Gamma(D + n - 2)}{2\Gamma(D)\Gamma(n + 2)}$

to be

$$i\nabla_i = L_i + \tfrac{1}{2}S_i. \tag{D.16}$$

The Laplacian can now be written as a sum of three angular momentum invariants,

$$-\nabla^2 = -4\nabla_i\nabla_i = 2J^2 + 2L^2 - S^2. \tag{D.17}$$

Tensor harmonics are constructed from the scalar harmonics by combining representations. A basis of scalar harmonics with angular momentum j is represented by rotation coefficients,

$$Q^{jmm'} = [j]^{1/2}(-1)^{m'}D^{jm'}_{m}(x) \tag{D.18}$$

where $[j] = 2j + 1$ and $x \in \mathrm{S}^3$. The angular momenta sum to give scalar eigenvalues $4j(j+1)$. These values are in agreement with the earlier polynomial constructions since j is a half-integer, and therefore $n = 2j$.

Spinor eigenfunctions can be obtained using

$$i\gamma^i\nabla_i = -4iS \cdot \nabla = 2L^2 - 2J^2. \tag{D.19}$$

Adding the **L** and **S** states using Clebsch–Gordan coefficients gives

$$\mathbf{Q}^{jj_1mm'} = \sum_{m_1,m_2} \langle j_1 m_1 \tfrac{1}{2} m_2 | jm \rangle Q^{j_1 m_1 m'}\mathbf{E}_S^{m_2} \tag{D.20}$$

where $\mathbf{E}_S$ is a spin basis, and the eigenvalues $\lambda^S = \pm(j_1 + j + 1)$. This agrees with table D.4, given $n = 2j_1$ and $(i\gamma \cdot \nabla)^2 = -\nabla^2 + D(D - 1)/4$.

References

Albrecht A. and Steinhart P. J. (1982) "Cosmology for grand unified theories with radiatively induced symmetry breaking" *Phys. Rev. Lett.* **48** 1220.

Albrecht A., Steinhart P. J., Turner M. and Wilczek, F. (1982) *Phys. Rev. Lett.* **48** 1437.

Arnowitt R., Deser S. and Misner C. W. (1962) in *Gravitation: An Introduction to Current Research,* ed. L. Witten. Wiley: New York.

Bardeen J. Carter B. and Hawking S. W., (1973) "The four laws of black hole mechanics" *Commun. Math. Phys.* **A31** 161.

Bardeen J. Steinhart P.J. and Turner M. (1983) "Spontaneous creation of almost scale free density perturbations in an inflationary universe" *Phys. Rev.* **D28** 679.

Beckenstein J. D., (1974) "Generalised second law of thermodynamics in black hole physics" *Phys. Rev.* **D9** 3292.

Birrell N. D. and Davies P. C. W. (1982) *Quantum Fields in Curved Space* Cambridge University Press: Cambridge.

Bond J. R. and Efstathiou G. (1984) *Astrophys. J.* **285** L45.

Branson T. B. and Gilkey P. B. (1990) "The asymptotics of the Laplacian on a manifold with boundary" *Commun. Part. Diff. Eqs.* **15** 245.

Brown J. D. and York J. W. (1993a) "Quasilocal energy and conserved charges derived from the gravitational action" *Phys. Rev.* **D47** 1407.

Brown J. D. and York J. W. (1993b) "Microcanonical functional integral for the gravitational field" *Phys. Rev.* **D47** 1420.

Callan L. and Coleman S. (1977) "Fate of the false vacuum II: First quantum corrections" *Phys. Rev.* **D16** 1762.

Carr B. J. (1976) "Some cosmological consequences of primordial black hole evaporations" *Astrophys. J.* **206** 8.

Carter B. (1973) in *Black Holes,* ed. C. DeWitt and B.S. DeWitt. Gordon and Breach: New York.

Chambers C. M. and Moss I. G. (1994) "Stability of the Cauchy horizon in Kerr–de Sitter spacetimes" *Class. Quantum Grav.* **11** 1035.

Chambers C. M. and Moss I. G. (1994) "Cosmological no-hair theorem" *Phys. Rev.* **D73** 617.

Chandrasekhar S. (1983) *The mathematical theory of black holes* Cambridge University Press: Cambridge, England.

Coleman S. (1977) "Fate of the false vacuum: Semiclassical theory" *Phys. Rev.* **D15** 2929.

Coleman S. (1985) *Aspects of Symmetry.* Cambridge University Press: Cambridge.

Collins C. B. and Hawking S. W. (1973) *Astrophys. J.* **180** 317.

Cornwall J. M., Jackiw R. and Tomboulis E. (1974) *Phys. Rev.* **D10** 2428.

DeWitt B. S. (1964) in *Relativity, Groups and Topology at the 1963 Les Houches School*, ed. B. S. DeWitt and C. DeWitt. Gordon and Breach: New York.

DeWitt B. S. (1967a) "Quantum theory of gravity I: The canonical theory" *Phys. Rev.* **D160** 1113.

DeWitt B. S. (1967b) "Quantum theory of gravity II: The manifestly covariant theory" *Phys. Rev.* **D162** 1195.

DeWitt B. S. (1967c) "Quantum theory of gravity III: Applications of the covariant theory" *Phys. Rev.* **D162** 1239.

DeWitt B. S. (1975) "Quantum field theory in curved spacetime" *Phys. Rep.* **C19** 295.

Dolan L. and Jackiw R. (1974) "Symmetry behaviour at finite temperature" *Phys. Rev.* **D9** 2904 and 3320

Dowker J. S. and Critchley R. (1976) *Phys. Rev.* **D13** 3224.

Esposito G. (1994) *Quantum Gravity, Quantum Cosmology and Lorentzian Geometries*. Springer-Verlag: Berlin.

Faber S. and Gallagher J. (1979) *Annu. Rev. Astron. Astrophys.* **17** 135

Feynman R. P. (1948) "Spacetime approach to non–relatavistic quantum mechanics" *Rev. Mod. Phys.* **A20** 367.

Feynman R. P. (1963) "Quantum theory of gravitation" *Acta Phys. Polon.* **XXIV** 697.

Feynman R. P. and Hibbs A. (1965) *Quantum Mechanics and Path Integrals*. McGraw–Hill: New York.

Gell-Mann M. and Hartle J. B. (1990) in *Complexity, Entropy and the Physics of Information*, ed. W. H. Zurek. Addison–Wesley: Reading.

Georgi H. and Glashow S. L. (1974) "Unity of all elementary-particle forces" *Phys. Rev. Lett.* **32** 438.

Gibbons G. W. (1975) *Commun. Math. Phys.* **A44** 245.

Gibbons G. W. and Hartle J. B. (1990) "Real tunnelling geometries and the large-scale topology of the universe" *Phys. Rev.* **D42** 2458.

Gibbons G. W. and Hawking S. W. (1977) "Cosmological event horizons, thermodynamics and particle creation" *Phys. Rev.* **D15** 2752.

Gibbons G. W. and Hawking S. W. (1983) *The Very Early Universe*. Cambridge University Press: Cambridge.

Gilkey P. B. (1984) *Invariance Theory, the Heat Equation and the Atiya–Singer Index Theorem* Publish or Perish: Wilmington.

Glashow S. (1961) *Nucl. Phys.* **B22** 579.

Goldstone J., Salam A. and Weinberg S. (1962) *Phys. Rev.* **127** 965.

Griffiths R. B. (1984) *J. Stat. Phys.* **36** 219.

Guth A. H. (1981) *Phys. Rev.* **D23** 347.

Guth A. H. and Pi S. Y. (1982) "Fluctuations in the new inflationary universe" *Phys. Rev. Lett.* **49** 1110.

Halliwell J. J. (1989) *Phys. Rev.* **D39** 2912.

Halliwell J. J. and Hartle J. B. (1990) "Integration contours for the no-boundary wave function of the universe" *Phys. Rev.* **D41** 1815.

Halliwell J. J. and Hawking S. W. (1985) "Origin of structure in the universe" *Phys. Rev.* **D31** 1777.

Hartle J. B. and Hawking S. W. (1976) "Path integral derivation of black hole radiance" *Phys. Rev.* **D13** 2188.

Hartle J. B. and Hawking S. W. (1983) "Wave function of the universe" *Phys. Rev.* **D28** 2960.

Hawking S. W. (1974) "Do black holes explode?" *Nature* **248** 30.

Hawking S. W. (1982) in *Astrophysical Cosmology,* ed. H. A. Brück, G. V. Coyne and M. S. Longair. Pontifica Acaemia Scientarium: Vatican City.

Hawking S. W. (1987) "Quantum coherence down the wormhole" *Phys. Lett.* **195B** 337.

Hawking S. W. and Ellis G. F. R. (1973) *The Large Scale Structure of Spacetime.* Cambridge University Press: Cambridge.

Hawking S. W. and Israel W. (1979) *General Relativity: An Einstein Centenary Survey.* Cambridge University Press: Cambridge.

Hawking S. W. and Moss I. G. (1982) "Supercooled phase transitions in the very early universe" *Phys. Lett.* **110B** 35.

Henneaux M. and Teitelboim C. (1992) *Quantisation of Gauge Systems* Princeton University Press: Princeton, NJ.

Isham C. and Kuchar K. V. "Representations of spacetime diffeomorphisms" *Ann. Phys.* **A164** 288 and 316.

Israel W. (1967) *Phys. Rev.* **D164** 1176.

Jackiw R. (1980) *Rev. Mod. Phys.* **A52** 661.

Kiefer C. (1987) *Class. Quantum Grav.* 4 1369.

Kirzhnits D. A. and Linde A. D. (1976) "Symmetry behaviour in gauge theories" *Ann. Phys.* **A101** 195.

Linde A. D. (1982) *Phys. Lett.* **129B** 177.

Linde A. D. (1984) *Rep. Prog. Phys.* **A47** 925.

Linde A. D. (1990a) *Particle Physics and Inflationary Cosmology.* Harwood Academic: Chur, Switzerland.

Linde A. D. (1990b) *Inflation and Quantum Cosmology.* Academic Press: San Diego.

Maeda K., Sato K., Sasaki M. and Kodama H. (1982) *Phys. Lett.* **108B** 98.

Mather J. C. et al. (1993) COBE preprint.

McKean H. P. and Singer I. M. (1967) *J. Diff. Geo.* 1 43.

Mellor F. and Moss I.G. (1990) *Phys. Rev.* **D41** 403.

Misner C. W., Thorne K. and Wheeler J. A. (1973) *Gravitation.* W. H. Freeman: San Fransisco.

Moss I. G. (1985) "Black-hole bubbles" *Phys. Rev.* **D32** 1333.

Moss I. G. (1989) "Boundary terms in the heat kernel expansion" *Class. Quantum Grav.* 6 759.

Moss I. G. and Dowker J. S. (1989) "The correct B_4 coefficient" *Phys. Lett.* **229B** 261.

Moss I. G. and Wright W. (1983) "Wave function of the inflationary universe" *Phys. Rev.* **D29** 1067.

Omnés R. (1988) *J. Stat. Phys.* **53** 893.

Page D. N. and Hawking S. W. (1976) *Astrophys. J.* **206** 1.

Page D. N. (1978) *Phys. Lett.* **79B** 235.

Page D. N. (1985) *Phys. Rev.* **D32** 2496.

Peebles P. J. E. (1980) *The Large-Scale Structure of the Universe.* Princeton U. Press: Princeton N. J.

Penrose R. and Rindler W. (1984) *Spinors and Spacetime* (2 vols). Cambridge University Press: Cambridge.

Pierce M. J., Welch D. L., McLure R. D., van den Berg S., Racine R. and Stetson P. B. (1994) *Nature* **371** 385.

Price R. H. (1972) *Phys. Rev.* **D5** 2439.

Regge T. and Wheeler J. A. (1957), *Phys. Rev.* **D108** 1063.

Ruffini R and Wheeler J. A. (1971) "Relatavistic cosmology and space platforms" in *Proceedings of Conference on Space Physics.* ESRO: Paris.

Salam S. (1968) in *Elementary Particle Theory: Relativistic Groups and Analyticity*, ed N. Svartholm. Almqvist and Wiksell: Stockholm.

Smoot G. F. et al, (1992) *Astrophys. J.* **396** L45.

Starobinskii A. A., (1973) *Sov. Phys. JETP* **37** 28.

Starobinskii A. A. (1980) "A new type of cosmological model without singularity" *Phys. Lett.* **91B** 99.

Tanvir N. R., Shanks T., Ferguson H. C. and Robinson D. R. T. (1995) "Determination of the Hubble constant from observations of Cepheid variables in the galaxy M96" *Nature* **377** 27

Teukolsky S. A. (1972) "Rotating black holes: Separable wave equations for gravitational and electromagnetic perturbations" *Phys. Rev. Lett.* **29** 1114.

Trimble V. (1987) *Annu. Rev. Astron. Astrophys.* **25** 425.

Unruh W. G. (1976) "Notes on black hole evaporation" *Phys. Rev.* **D14** 870.

Vassilevich D. V. (1995) "Vector fields on a disk with mixed boundary conditions" *J. Math. Phys.* (to appear).

Vilenkin A. (1982) "Creation of universes from nothing" *Phys. Lett.* **117B** 25.

Vilenkin A. (1984) *Phys. Rev.* **D30** 509.

Vilenkin A. (1988) *Phys. Rev.* **D37** 888.

Vilenkin A. and Shellard E. P. S. (1995) *Cosmic Strings and Other Topologcal Defects.* Cambridge University Press: Cambridge.

Wald R. M. (1993) "Black hole entropy is the Noether charge" *Phys. Rev.* **D48** R3427.

Wald R. M. (1994) *Quantum Field Theory in Curved Spacetime and Black Hole Thermodynamics.* Chicago university Press: Chicago.

Walker T. P. et al. (1991) *Astrophys. J.* **376** 51.

Weinberg S. (1967) *Phys. Rev. Lett.* **19** 1264.

Weinberg S. (1972) *Gravitation and Cosmology.* Wiley: New York.

Weinberg S. (1974) "Gauge and global symmetries at high temperature" *Phys. Rev.* **D9** 3357

Zel'dovich, Ya. B. (1972) *Mon. Not. R. Astron. Soc.* **A160** 1P.

Index